Scanning and Image Processing for the PC

Frank Baeseler • Bruce Bovill

Scanning and Image Processing for the PC

McGraw-Hill Book Company

London • New York • St Louis • San Francisco • Auckland •
Bogotá • Caracas • Lisbon • Madrid • Mexico • Milan •
Montreal • New Delhi • Panama • Paris • San Juan •
São Paulo • Singapore • Sydney • Tokyo • Toronto

Published by
McGRAW-HILL Book Company Europe
Shoppenhangers Road, Maidenhead, Berkshire, SL6 2QL, England
Telephone 0628 23432
Fax 0628 770224

British Library Cataloguing in Publication Data
Baeseler, Frank
  Scanning and Image Processing for the PC
  I. Title   II. Bovill, Bruce
  006.6

  ISBN 0–07–707819–5

Library of Congress Cataloging-in-Publication Data
Baeseler, Frank
    Scanning and Image Processing for the PC / Frank Baeseler, Bruce
Bovill
      p.   cm.
    Includes index
    ISBN 0–07–707819–5
    1. Optical scanners.   2. Image processing.   3. Microcomputers.
  I. Bovill, Bruce   II. Title.
  TK7882.S3B33   1993
  004.6`2—dc20   93–31120
                                                          CIP

Typeset by Pentacor PLC, High Wycombe, Bucks
and printed and bound in Great Britain, at the University Press, Cambridge

Contents

Chapter 4
Originals for Scanning   **38**

Chapter 5
Hand-held Scanner: ScanMan 256   **46**

Chapter 6
Flatbed Scanner: HP ScanJet IIp   **65**

Chapter 7
Resolution, File Size and File Formats 96

Chapter 8
Scanning with Colour 104

Chapter 9
Programs for Image Processing 112

Preface

This book describes what you need to know when you want to choose—and use—an image scanner and it provides an overview of the kinds of images which lend themselves to being scanned and manipulated. Along the way, we will give hints as to how best to use a scanner, and help in the day-to-day process of scanning. If you are new to scanning, the book will provide an invaluable insight into the results that may be achieved. If you already own or have access to a scanner, we will help you to maximize the effectiveness of your usage of this rapidly developing technology.

The idea for a book on scanning came to us over mugs of coffee one day. We both work with personal computers on a daily basis, and are both involved more with the presentation of information than with pure numeric computing. We had used each of the systems which make up the majority of the personal computer market—those based on the Intel processor chips (IBM and IBM-compatibles), and those which use the Motorola chips (Macintoshes), though most of our work involved IBM systems. We were fascinated by the new capabilities which a scanner could open up, but realized very quickly that there was very little help available to enable us to make sensible decisions before spending money. We knew very little about the kinds of problems we might meet, or the facilities which would be available to us. We also realized that, if this was true for us, then it would be more of a problem to many potential scanner users who might not have our contacts in the hardware and software industries, and so we decided to write a guide for the user of the IBM PC. Scanners and powerful image processing capabilities have been available for the Macintosh systems for several years, but it is only lately that the IBM user has had the benefit of integrated facilities. Frequently the IBM PC has been underrated as an image processing tool, and we hope that this book will begin to redress the balance. In our daily usage we have found a range of scanners and associated software which provide facilities which are no less easy or effective in use than their Macintosh equivalents. The advent of new function-rich applications software for Microsoft Windows 3.1 will undoubtedly improve the situation still further

We have tried to avoid using too many technical terms and acronyms but some have necessarily crept in. If they are not explained where they are used, they will be found in the Glossary.

The reader need not worry how many permutations exist of hardware which can provide scanning capabilities, since most current models do basically the same job. The system interfaces may vary, for example, in hardware, there will be differences between scanner cards which are for a MicroChannel Architecture systems such as the IBM PS/2 and those using the Industry Standard Architecture (ISA or AT) I/O bus. In software, there will be differences between scanning applications which use Microsoft Windows and those which do not. An image to be scanned is the same image, no matter which scanner is used, and an image file is likely to conform to an industry standard format, such as Tagged Image Format (TIFF). The major differences will be found in the software which drives the scanner, in some special formats for storing the image files, in the enhancing of an image or in printing it out, and in the sizes of RAM, fixed disk or floppy disk which are needed by individual programs.

Scanning and image manipulation are an integral part of the scanning process, and this book would be incomplete if we were not to discuss different software available to the IBM user. We have found certain software to be particularly effective used with any scanner, and we have also used different types of scanner on a daily basis. We try to ensure that you will get the relevant information to help you choose the right combination of hardware and software. Without the right software, any computer is, of course, just so much plastic, metal and silicon and this is no less true of a PC used for scanning. Only with the right system software and programs for scanning and enhancing graphics will we find the scanner to be an indispensable tool in our working environment. Fortunately, as time goes by and scanning as an application matures, the standard packages that we can buy include most of the functions that we may need.

We reckon that scanners will soon become as essential to the user of a PC as a mouse and printer have become. For several years, enhancements of the presentation of the printed word in the electronic publishing revolution have taken the limelight; with the addition of effective image scanning the revolution in everyday document presentation can take another gigantic step forward.

Our aim in this book is to show how to scan, manipulate and optimize pictures, but before we begin we would like to acknowledge the invaluable help given by the different companies which supported us in writing the book: Aldus Software GmbH; Microsoft GmbH, DTP Partner Hamburg, LOGI GmbH and LOGI (UK) Ltd, Makrotronic AG, Microtek Europe GmbH, CPI, Hewlett-Packard Ltd, 2020 Technology Ltd, and the University of London Computer Centre. With their help it has been possible to give the reader a snapshot of the current market, and a series of recommendations for both hardware and software parts of the scanning process.

Frank Baeseler   Bruce Bovill

Chapter 1
Introduction

When the IBM Personal Computer was introduced in 1981 few of us could foresee the technical development that has taken place since then. The first IBM PC came with only 64K RAM, without a fixed disk and with no graphics. Within a few years it grew to be a computer system which, on a desktop, had the power of a mainframe which just a short time before, required a large amount of floorspace in a specially conditioned room. Even then, one could not have dreamed of the things which are not only possible using today's PC, but quite normal. Until quite recently, the hardware and software necessary to provide the new levels of functionality were neither available as complete systems, nor as configurable components.

Since 1981 there has been tremendous progress with the microprocessor, the heart of any PC. Someone who until recently would have purchased a system based on the Intel 80286 processor now finds the 80386 or 80386SX to be the logical choice, and can see faster and faster 80486s looming on the horizon, and with the next generation, Pentium, processor about to be launched. Where it was considered a great advance to have fixed disks of 5 and 10 megabytes in the early days of the PC, it is now considered quite normal to have capacities counted in hundreds of megabytes with much faster access times. Disk caches are now in common use.

Flexible media have also developed: the original storage capacity of 180K and 360K has increased to 1.44Mb with 3.5 inch floppy diskettes, and the 2.88Mb floppy diskette is not far off shipping in quantity. The majority of computers purchased today are supplied with at least one megabyte of RAM, and this can be expanded (and used) up to 32Mb in the new generation of systems without any problems.

THE INFLUENCE OF GRAPHICS

Most people today expect a personal computer to be able to produce and display quite complex graphics. The original character-oriented IBM PC has developed into a graphics-oriented machine via the Hercules, CGA and EGA graphics standards to the current VGA, Super VGA and XGA.

The general availability of high resolution graphics has influenced software developments so that user friendly interfaces such as Microsoft Windows 3.1 are becoming the norm. The mouse has become essential as a pointing and input device for the graphical user interfaces such as GEM, Windows and OS/2 Presentation Manager.

THE DEVELOPMENT OF PRINTER TECHNOLOGY

Over the same period, printing technology has also developed by leaps and bounds. Early dot matrix printers used a 9-pin head to produce text and graphics. For both of these kinds of hard copy, 9-pin printers produce results which are really only acceptable for drafts. Whenever a high quality text printout was needed people used daisy wheel printers or electric typewriters with a computer interface. More recently, dot matrix printers have moved to the 24-pin standard (Fig. 1.1), and 48-pin printers are now commercially available. Some of these print in colour.

The main problem with dot matrix printers is the noise they generate while printing and, although printing speeds have been improved dramatically, the problem of noise has never been fully addressed.

In parallel with these developments has been the introduction of new technology devices, such as ink jet printers and high resolution laser printers (both normally in black and white, but also available in colour). Ink jet printers usually have the drawback that their ink is not fixed in the same way as that for impact or laser printers: it may run if exposed to moisture. The most commonly used laser printers can print at 300 dots per inch (dpi) for text and graphics (Fig. 1.2). Some laser printers print at 400 or even 600 dpi. Colour laser printers are now available but will remain very expensive until they penetrate the market in quantity.

The main problems affecting the use of high resolution colour graphics for output with anything other than a plotter of some kind, are those of price and speed when compared with equivalent monochrome printers.

THE PROBLEMS OF DUPLICATION

Producing multiple coloured hard copies from a colour output device is an expensive and time-consuming business. When ten or more copies are needed, time considerations are likely to dictate that the services of a print shop are used—with subsequent external costs and a lot of additional time. It is a different story for black and white as we can use normal photocopying machines for fast production of small to medium quantities. The main advantage in this kind of printing is the high penetration of the necessary hardware in companies and institutions. The situation will change with time,

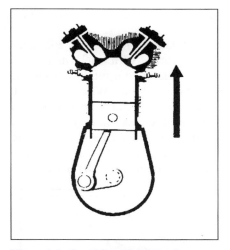

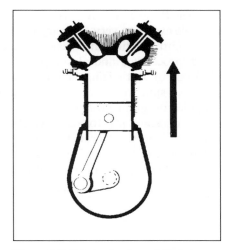

Figure 1.1 Graphics printout using a 24-pin matrix printer

Figure 1.2 Printout using a 300 dpi PostScript laser printer

and we will see a rapid growth in colour hard copy over the next few years—the acceptance of high quality, colour hard copy in place of the current monochrome, lower resolution, will be similar to the evolution from the 64K RAM PC to an 80486 power system.

COMPETITION BETWEEN SYSTEMS

Competition stimulates technological development, and it has been the Macintosh systems which have pointed the way for the IBM developers. The user of the earliest Macintosh had graphical capabilities denied the IBM and IBM-compatible systems user. The advent of the Macintosh and its innovative interface, following the unexpected success of the IBM PC, led developers to try to catch up and even overtake the Apple systems.

IBM has really only just caught up to the Macintosh functionality using appropriate application software together with a similar kind of user interface (Presentation Manager with OS/2 and Windows 3.1 with DOS). Until recently IBM systems have only been ahead in market share, not in technical innovation.

PRESENTATION BY COMPUTER

The effect of the PC revolution is quite clearly visible. Documents such as internal memos, annual company reports or presentation material for conferences are only taken seriously if they are produced in a professional manner. Rescue, however, is at hand—a high degree of professionalism is now possible even in a very ordinary PC set-up.

The introduction and general acceptance of laser printers and enhanced word processing and desktop publishing programs, with their associated abilities for optimizing layout and producing quality output has been the principal spur to the improvement in presentation material. Where once each graphics program produced results only in its native format, now there is a range of standard formats which may be used either directly by other programs, or via specialized format translation programs. Page layout and enhanced word processing programs can now accept a wide range of these formats, making the inclusion of graphics with text a much simpler matter.

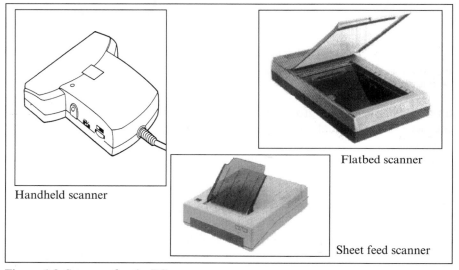

Figure. 1.3 Scanners for the PC

IMAGES AND SCANNERS

In the short time that PCs have been involved in the publishing and presentation process we have become used to seeing graphs, bar charts and pie charts incorporated within typeset documents, but the ability to incorporate photographs and other external images has been missing. If we look at any newspaper or magazine, we see photographs, maps and diagrams. These are the final elements we want to add to our publishing repertoire, and they are now possible with the help of a scanner (hardware) and the appropriate programs (software) to control the scanner and to manipulate the scanned originals (Fig. 1.4). We will give more details of all aspects of scanning throughout this book.

Scanners in various forms have been available for many years, but within the PC environment they have been, to some extent, a solution that is in search of a problem (Fig. 1.3). They could only really find their market once the other building blocks for professional presentation had been laid. We can summarize those blocks as the desktop publishing boom. It has taken a very short time for the potential of this market to be appreciated, and for a variety of different 'image capture devices'—some of which are low in price—to come to the marketplace.

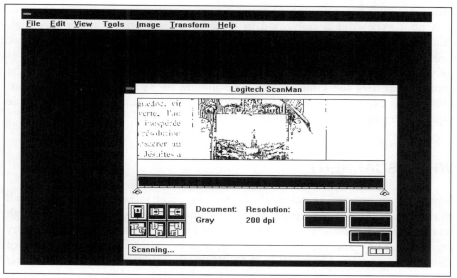

Figure 1.4 Software to control the scanner

THE RIGHT DECISION

What can a scanner do? What do we have to take into consideration when buying one? Does a low-priced scanner have many less features than an expensive one? How simple or complicated is it to set up and is it difficult to use the software? The chapters that follow will provide answers to these and other questions.

QUALITY THAT SELLS

The personal computer is now very widely used. It is a major tool for accounting, inventory, sales planning or development, statistical analysis, as well as for word processing and documentation. Systems are used within organizations for pure administrative work, as well as for the production of externally targeted material.

In today's competitive world the successful presentation of a product, or of the company itself, can be the key to success. Now we can produce in-house most of the presentation material that we need. Using a desktop or electronic publishing system has, within a very short space of time, radically changed the way in which many company departments work. Fine art work for printed material can be generated internally and need not be handed over to external service people such as designers and typesetters. The main advantages of this are the shortening of long production times, additional flexibility and the ability to make last-minute adjustments while having full control over the final product.

A phototypesetter is only used in place of a laser printer when very high quality art work is needed. The desktop publisher can hand over a floppy disk containing the page layout data. With a professional typesetting machine it is only a matter of an hour to create high resolution printing films. So, the professional uses the same system environment as we may typically use, but invests in expensive professional output devices, such as a Linotronic typesetter rather than a PostScript laser printer.

WHY A SCANNER?

Scanners are not yet as common as mice or printers but this situation will change dramatically over time. Using a scanner enables us to produce high quality printed documents. We can transfer in-house product shots, company logos, diagrams and the like directly into sales material, and this does not restrict us to printed matter (Fig. 1.5). With appropriate software we can use scanned images—even coloured ones—for electronic presentations using the computer. Lastly, we can change and improve the scanned image. We will cover this in detail later.

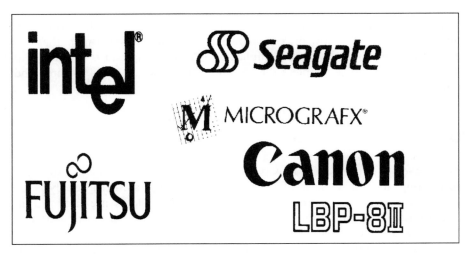

Figure 1.5 Scanned company logos (300 dpi) without any improvement

Chapter 2
Technology and Installation

THE PRINCIPLE

A scanner is a cross between an electronic camera and a photocopier for capturing and enhancing printed originals (for example, line art and photographs) using a computer. Preparation of originals for use with a computer is often called digitizing. In this process an image is broken down into its individual picture elements or pixels. A scanner is usually delivered with software which enables originals to be read, or 'scanned'.

In contrast to a real camera most scanners can only capture flat, two-dimensional images, and in this a scanner is similar to a photocopier. Externally, most scanners (with the exception of hand-held and camera ones) even look like slimmed-down photocopiers, and this is no coincidence as many of the functions are the same. Both types of device scan an original; the one to reproduce it on paper, the other to digitize it.

While a photocopier produces one or more copies immediately after scanning (though we can adjust the image via contrast controls), a scanner scans the original with an opto-electronic component which has many electronic sensors set in a line. These sensors may be photodiodes or charge coupled devices (CCDs). The CCD is an optical chip with many small sensors on the surface. These sensors convert different light intensities into defined voltage values. Similar CCDs arranged in a matrix may also be found in most video camcorders.

During the scanning process the object being scanned is illuminated by a light source in the form of a stripe or a circle (Fig. 2.1). The white parts of the original reflect more light than the dark ones. These different light reflections are sent by an optical system using mirrors or fibre optics to the CCDs, which generate a voltage which is relative to the intensity of the reflected light. This analogue voltage is digitized with the help of an electronic component called an Analogue/Digital (A/D) converter. The computer can then read the information. We will cover this in further depth in the following chapters, especially with reference to bit depth and grey steps.

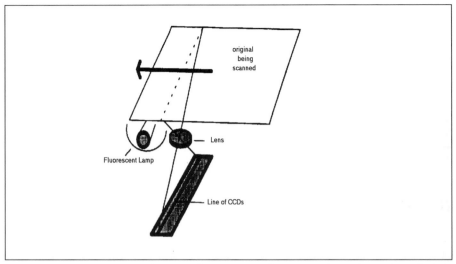

Figure 2.1 Scanner with a CCD scan line (original, fluorescent lamp, lens, CCD-line)

SCANNER AND PRINTER

Unlike camcorders, which can recognize a wide variety of tones and colours, most scanners and photocopiers are dominated by just black and white. Scanners differ in another way from both photocopiers and electronic cameras: an original is scanned dot by dot, in much the same way as a laser printer prints. While a laser printer creates an image by combining many single dots, a scanner converts an image into those single dots. As long as there are sufficient dots available in the scanned version of an image, there will be very little difference between the scanned and original versions. In desktop publishing applications in particular, we can assume that most scanners and laser printers will be compatible in terms of resolution.

SCANNER CONSTRUCTION

Different types of scanner look quite different from each other, and originals are presented to them in various ways. In the main, scanners fall into these categories:

- Hand-held
- Those where an original is fed into a roller-driven transport mechanism to be scanned
- Flatbed and camera scanners.

Digitizers and frame grabbers are a more specialized area of the scanning market which we will not cover.

The majority of professional scanner users today use flatbed scanners. The hand-held scanner is a sensible choice for the beginner or for the person who will only ever need to scan small originals. The main disadvantage is its compact size and the care which needs to be taken to produce an accurate rendering of the original. Camera scanners are only available at present from Linotype and Chinon.

Scanners are constructed with two basic differences: either the detector moves over a fixed original or an original is pulled in and led over the detector. This last method, which is used where originals are fed into a roller-driven device, frequently causes imprecise presentation of the original to the detector. This may lead to repeated attempts to get a good scan. In addition, small originals, such as business cards, stamps and Polaroid photographs cannot be handled by the transport rollers of the scanner.

The flatbed scanner has no such problems and it has many advantages. For example, originals of varying thicknesses, sizes and forms may be scanned without problems. Results are more easily guaranteed as no manipulation of the original is made by mechanical parts which will wear over time. A static original enables the scanner to make a first quick pass at low resolution, so we can delineate the precise area that we want to scan. A slower, high resolution scan of a reduced area may then be made. These factors lead to more accurate and faster scanning.

For those who need to have originals fed automatically, and this is typically necessary where OCR (Optical Character Recognition) scanning is necessary, optional sheet feeders for continous scanning of a number of originals are available. The disadvantage is that automatic sheet feed has some of the problems of the roller-driven scanner in the accuracy of registration of the sheets being fed. Fortunately, it is unusual to need to use a sheet feeder to scan multiple graphic images, as they rarely will appear in the same position on consecutive pages, whereas text normally will.

Roller scanners work with best effect when they are presented with an original which is the maximum size they can accept. This will normally be A4. In this way, the best accuracy may be achieved in feeding the original. A tip to help overcome problems encountered when scanning originals which are smaller than A4 is to fix small or irregular sized originals within a transparent, wider envelope. This will tend to reduce the precision, but will make the process easier, or indeed, possible. Scanning of a thick original is only possible by photocopying the original book or brochure, or by cutting out the parts that are needed.

The type of scanner one person will choose depends on the kind of work that will be expected of it. Certainly where large tracts of text are to be read, flatbed scanners with sheet feeders have an advantage. This is because they can be used for uninterrupted automatic scanning of many pages of information that will be converted using OCR software into editable ASCII (unformatted) or DCA (formatted) text files.

Roller-fed scanner

Most roller-fed scanners can handle originals up to A4 size (portrait). A special paper path is used for aligning the original which then is pulled by motor-driven transport rollers over a 'window'. Under this will be found the light source, the lenses and the detector. This arrangement permits the scanner to be quite compact. The original is pulled over the sensor and not—as with the flatbed scanners—the sensor pulled over the original (Fig. 2.2). Therefore the basic size of a roller-fed scanner is often even smaller than the size of the original. A similar construction principle is used with telefax machines.

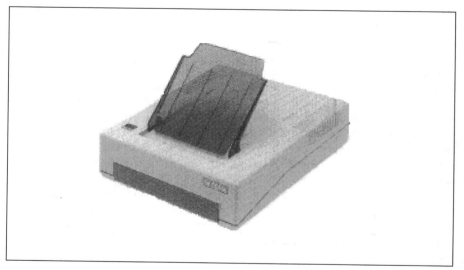

Figure 2.2 Roller-fed scanner

Flatbed scanner

Flatbed scanners look outwardly similar to compact copying machines or light boxes for viewing transparencies. Flatbed scanners are normally around 10 centimetres high and they are a little bit wider and longer than the maximum size of the original to be scanned (Fig. 2.3). The document is placed face down on a glass surface and is then pressed on the surface by a flexible cover which also protects the original from any light other than that used by the scanning mechanism. This is a familiar arrangement for users of copying machines. The next step is the automatic scanning of the original, where a motor moves either the complete sensor or—on some machines—an arrangement of mirrors. We have already mentioned the advantages of this construction principle.

Figure 2.3 Flatbed scanner

Camera scanner

Camera scanners look like photographic enlargers and work in the same way as a photographic camera (Fig. 2.4). The lenses likewise can be changed. An original is fixed on a platform while the scan device is placed in a movable box on a vertical support. The scan device is positioned and fixed in its final position according to the size of the original and the lens used before starting the scan procedure. The lens then projects the image of the original on to a sensor which contains a CCD matrix: the original will be digitized as a whole and transferred to the computer. The Truvel TZ-3 professional camera scanner is a typical example of this design and is used within complete desktop publishing systems like the one marketed by Linotype. This particular scanner has a resolution up to 900 dpi and creates up to 256 different grey steps.

Other camera scanners, such as the Chinon (which has a resolution up to 300 dpi), use a mirror system that scans the lens-projected images line by line. This information is transferred to the detector with the sensors and then digitized and sent to the computer. This means that the scan procedure takes some time and it is important that the image is not subjected to any vibration during the process. This is also true for other types of scanner. The original must also be evenly lit as an irregular light across it will cause an irregular scan to be made.

Figure 2.4 Camera scanner

Hand-held scanners

The original hand-held scanners could only be used with small originals or when scanning only small sections of a large original, owing to their relatively small scanning head. Recent developments mean that larger images can be scanned in strips, and then these can be stitched together by software to provide a larger scanned image. In principle hand-held scanners work in much the same way as flatbed scanners except that the entire scanner is rolled over the original, and not just the scanning head (Fig. 2.5).

Hand-held scanners take their power directly from the computer. This power is needed for the integrated light source (yellow/green or red light emitting diodes, LEDs) and some internal electronics. By rolling the scanner the original will be illuminated by the LEDs and the reflected light is then received by a sensor array. The image is then digitized in the same way as with the roller-fed and automatic feed scanners.

The maximum scan width for the original (or part of an original) differs for the available hand-held scanners from 60 to 100 millimetres and the maximum length from 250 and 280 millimetres. If we scan a long original at the highest resolution (400 dpi) the amount of free memory (RAM) in our computer becomes an important factor.

Later in this book we will learn in detail how to work with a hand-held scanner.

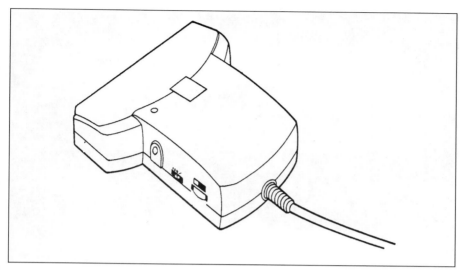

Figure 2.5 Hand-held scanner (Logitech ScanMan 256)

Digitizer or frame grabber

This particular technique for creating digitized pictures for the computer does not directly relate to scanners. We will only mention these devices in passing, to give us a more complete picture of all available imaging means.

A frame grabber provides a connection between a PC and different kinds of video sources and is used to translate the image from a video camera or a video recorder for the PC. As with scanners these fall within the general category of 'image processing'. Any object screened by a video camera, or single images (frames) within a video clip, may be digitized by the frame grabber, stored in its separate memory and then transferred to the computer. All further steps, such as picture manipulation or saving in one of the standard graphic file formats, are the same as with scanners.

A frame grabber normally consists of a PC adaptor card that captures images or video signals either in black and white or in colour. The capture can be very fast (1/50 second) where there is any movement, or within a couple of seconds for still images. Both systems work with different resolutions and grey or colour steps.

An extension of the frame grabber is the PC video printer (Fig. 2.6). The analogue video signals may also be digitized and stored by this device but then the information is printed in black and white or in colour on to thermal paper. Video printers can also process signals from the video adaptor on the PC. The resulting coloured screen shots will take some minutes to print.

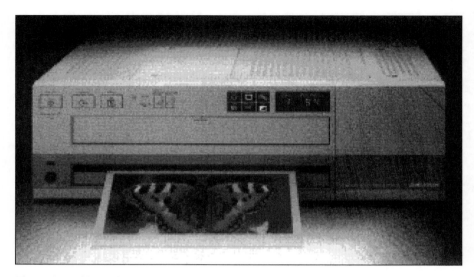

Figure 2.6 Video printer

CONNECTING TO A PC

Until now we have looked at the different scanner designs, but not how a scanner will be connected to a PC. A first glance through the manuals, instruction or installation books of the different scanner manufacturers can easily frighten the normal user. The combinations of hardware and software may seem very complicated. Let us look in more detail at the various bits and pieces.

Interface Card

Virtually all scanners are connected to their computer by a special interface card (Fig. 2.7). Typically, most scanner manufacturers will expect there to be enough room in the IBM PC or compatible for this additional card to be fitted. This is not always the case, and it could seem to be a serious problem. If the computer is crammed with cards, we will need to remove one of the existing cards in order to install the scanner interface card. Alternatively, we can try to combine two of the existing cards into one card, by buying a multi-function card. If this is not possible, then buying a new PC with space for additional cards might be the answer. More and more systems these days have functions integrated into their motherboards, functions that once needed to be held as individual cards.

Another solution could be a scanner with a serial interface, for example, the Chinon camera scanner, the Epson GT-1000/GT-4000 and the Sharp JX-100. In such a case there is no problem connecting the scanner with the serial port on the PC.

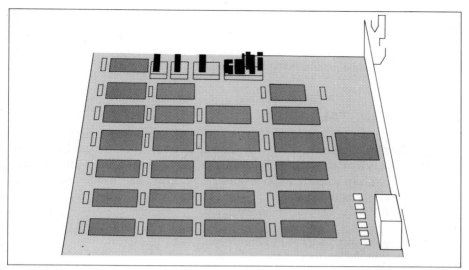

Figure 2.7 Interface card for a scanner

Another problem arises if we own a computer system with a MicroChannel (MCA) Interface (a physically different system needing special interface cards). At one time very few scanner manufacturers offered a MicroChannel card (for example, for the IBM PS/2 series, starting with model 50 or compatibles). If we have an MCA system, then it is essential that we select a scanner that has an appropriate controller card.

PREPARING FOR INSTALLATION

If our scanner is not to be connected and installed by a dealer we should first unpack the various bits and pieces—the scanner itself, an interface card and a connecting cable. Hand-held scanners have an integrated cable through which they take their power. All scanners need external power, so check for the power cord, unless it is a hand-held model. It is almost certain that the scanner will have been preset for the power voltage in our area. Check it anyway! So the hardware is all present and correct. Now we will look at the documentation and the floppy disks containing the installation and application programs. Usually, the scanner handbook will list the contents of the scanner package showing the standard accessories.

Floppy disks and software

Do check that all of the floppy disks containing the installation and application programs are there. A scanner without software is worthless, but even with a complete set of floppy disks we will be in trouble if the package does not contain the format of diskette that our computer can read. Fortunately, many personal computers these days have both 3.5 inch and 5.25 inch diskette drives as standard.

A SAMPLE INSTALLATION

Let's go through the installation of a scanner and the interface card using an IBM PC, XT, AT or compatible. Most scanners are installed in a similar way. Using the Logitech ScanMan 256 hand-held scanner, the installation will be performed in a series of steps:
Computer: IBM PC, XT, AT or compatibles:

1 Inserting the interface card in the PC.
2 Connecting the scanner to the PC.
3 Installing the software.
4 Testing the scanner.

There is a slight difference with IBM PS/2 systems. All we need to know is contained in the scanner documentation and these are the steps:
Computer: IBM PS/2 (model 50 upwards):

1 Inserting the interface card in the PC.
2 Connecting the scanner to the PC.
3 Configuring the computer with the reference disk.
4 Installing the software.
5 Testing the scanner.

Installing the interface card and connecting the scanner

First, we power off the computer. Next, we remove the cover from the computer's system unit. This will give us free access to the expansion slots. Then we choose a free slot and remove the appropriate cover at the back, so that we can insert our scanner interface board firmly and fix it with a screw. Now all we need to do is to plug the scanner cable into the port on the scanner interface card. Before we power the system back on, we should protect the delicate electronic and mechanical parts by replacing the system unit cover— but we do not secure it yet. First wait until we have been able to demonstrate that the installation has been successful, then secure the cover properly.

Installing the software

The ScanMan 256 is supplied with Windows-based software to drive the scanner and manipulate the images that result. Once we have successfully fitted the scanner card into the system unit of the computer, and attached the scanner, the installation program needs to be run. This will not only enable us to drive the scanner—it will also tell us whether the installation has been successful.

Before we can install the software however, we have to start up Windows. Once this has been successfully done, we can select the FILE menu of the Program Manager window and choose *run*. This opens a dialogue box, and we will be prompted to type in either A:install or B:install, according to which floppy drive is being used for the installation. Both formats of floppy diskette are included with the scanning package. It is a matter of moments to load the software—the install program gives us a running commentary on the process once we have told it where to store the software (Fig. 2.8).

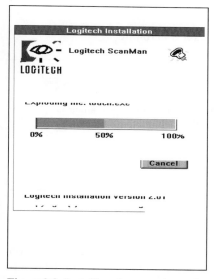

Figure 2.8 Installing the scanning software

Immediately the copying process has been completed, a new dialogue box, '*ScanMan Hardware Setup*' appears and invites us to test the scanner. It is certainly worth trying it right away, without attempting to change any of the settings (Fig. 2.9).

When we click on the *Test* button, the '*ScanMan Test Procedure*' dialogue box asks us to check whether the light on the scanner has been activated. If no light has appeared, we click on *No Light*, and the program tells us that no scanner activity has been detected (Fig. 2.10). This may simply mean we have not moved the scanner at all, or it may mean that we will need to change one or both of the settings.

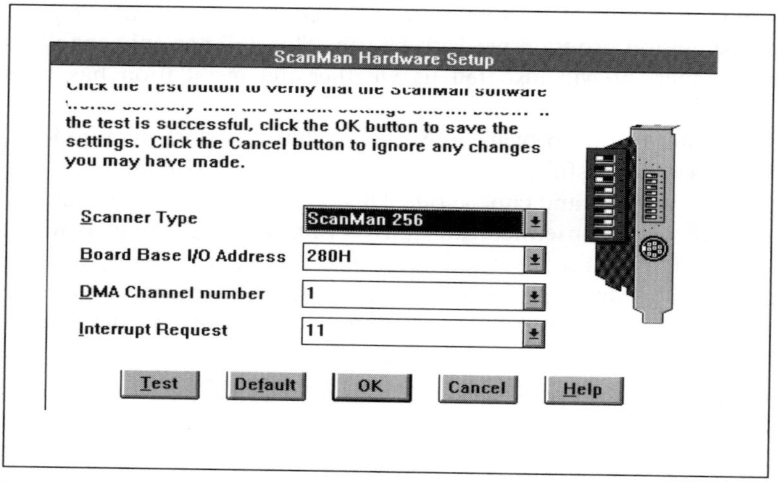

Figure 2.9 Co-ordinating the software with the scanner hardware

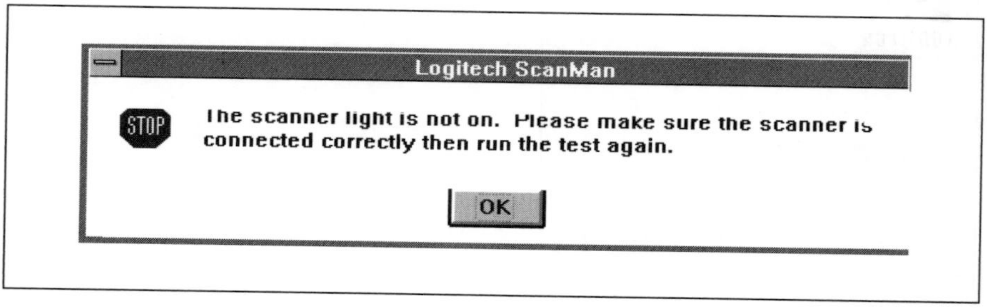

Figure 2.10 Scanner has not been installed correctly

If the light has appeared, then we need to press and release the scan button, and move the scanner so that it can scan a sample image. We don't have to move the scanner far before the installation program tells us that the scan test has been successful (Fig. 2.11).

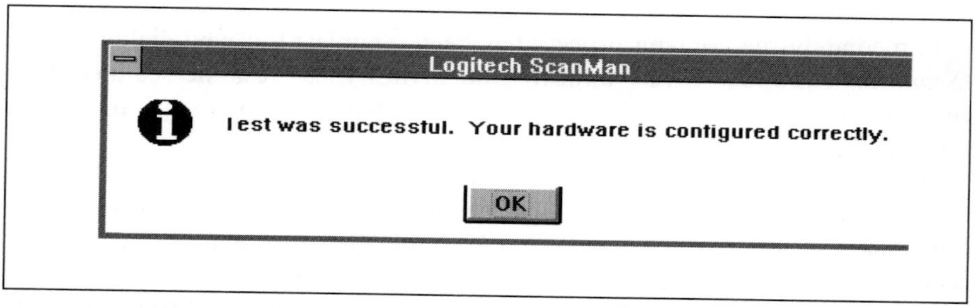

Figure 2.11 Installation program shows the scanner installation has been successful

We are then invited to restart Windows. This is so that all of the changes made in the installation will be reflected in the scanning we do. In many cases, no adjustment will need to be made to the scanning hardware. However, if any clashes are detected with the existing set-up of the system, then it will be necessary to adjust the DMA (Direct Memory Access) channel used or the interrupt request number. Help for this is contained in the scanner manual.

Making adjustments

If we have to adjust the *Base I/O Address*, then we have to do this first on the adaptor card. The base address enables the computer to recognize the interface card and to communicate with the scanner. The default address is 280H. When there are incompatibilities with other devices connected to the computer we have to choose from the alternative addresses: 2AOH, 33OH and 34OH.

Once we have changed the hardware base address, we must first of all re-boot the computer. Then start Windows up again, rerun the install program, and also change the base address setting within the software. It must agree with the dip switch setting to permit the scanner to be used.

The *Interrupt Request* may need changing. The operating system of any computer recognizes input devices by means of interrupts, which are sent via the PC bus from the input device to the microprocessor over channels IRQ2 to IRQ7 in the IBM PC. IRQ2 has the highest and IRQ7 the lowest priority. Expansion and interface cards must have different IRQs to avoid system conflicts. When two cards are configured for the same IRQ and are active at the same time the result will cause the system to hang, in which case the only solution is to shut down the computer, make an appropriate change and start it up again. The matching IRQs are, as for the addresses, set by means of jumpers on the interface card. Not all interrupts are available for this kind of use—some are reserved by the computer itself. Table 2.1 shows some examples of restrictions.

Table 2.1 Reserved IRQs

System	Reserved IRQ
IBM PC, XT	5
IBM PC, AT	2
IBM PS/2 (model 25/30)	5
IBM PS/2 (model 30/286)	2
COM1 (1. serial port)	4
COM1 (2. serial port)	3
Graphics card (EGA or VGA)	2
LPT1 (1. parallel port)	7
LPT2 (2. parallel port)	5

It is best to check the status of the IRQ settings for the particular system before embarking on setting up the interface card for the scanner. Normally the default setting IRQ3 set as default by Logitech should be used.

A DMA channel is used by the computer for direct access to memory. The settings in Table 2.2 are recommended for the computer systems shown:

Table 2.2 DMA channels

System	DMA channel
IBM AT	1 or 3
IBM PC, XT or PS/2 (25/30)	
without fixed disk	1 or 3
with a fixed disk	1

As channel 1 is set by default, and this will work in all cases, it is sensible to leave it set at that value. Changing it will require us to reset jumpers on the interface card. Once the installation has been successfully completed, we can leave the installation program and remove the floppy disk containing the program from the floppy drive. In future, each time the computer is powered on, it will be ready for us to start scanning. According to the objects we want to scan, we will then need to set some values within the scanning programs for resolution, scan width and time out.

This has been one example of the things which must be done when installing a particular scanner and gives a flavour of what needs to be done. This installation may differ in detail from the one we will ultimately need to do. Now we have finished the installation, we can start to explore what the scanner can do. A good starting point is the *Logitech ScanMan Software User's Guide*, which includes a tutorial on the basic use of the scanner. We won't need to load any additional software (at least for a while) because the process of loading the ScanMan scanning software will also have loaded Logitech's FotoTouch image editing software. Chapter 5 will describe in depth how to work with these programs.

If we have got to this stage successfully, we can turn off the computer and secure the system unit. Then we can start the computer again to produce our first scans. The installation of a scanner—the hardware and software set-up—is something that has to be done. But it should only need to be done once! Then we have a system with a whole new functional horizon, and we can bring almost any image into our computer to incorporate into documents, to enhance it—to do whatever we want to do—within the capabilities of our software.

Chapter 3
Resolution, Grey Shades and Screens

In this chapter we will describe some of the theoretical and practical basics for effective scanning. Although scanner manufacturers will stress the importance of their machines having the highest resolutions and the most grey scales as their main selling points, for most people a very highly specified system is not what is needed. This is especially true when scans are to be printed using a laser printer that has a maximum resolution of 300 dpi. A high number of greys is only needed when a scanned halftone is to be processed by an imagesetter such as a Linotronic 100/300. Screening is used to define the number of grey tones within the printed scan for a particular output device, and it is important whether a laser printer or a very high resolution imagesetter is being used for printing.

REPRESENTATION OF GREY TONES

Grey tones are represented differently for computer monitors and for printers. The computer can recognize two values for a scanned halftone—black and white—and special techniques are required to show and print greys which fall between the two.

Line art subjects, which only contain black and white, are represented using only two digital values, so there is no problem displaying scanned line art on a monitor and in printing, as only one bit is needed for each image dot or pixel. This single bit sets the value for a black or white pixel. Halftones such as black and white photographs are more complex, as they feature many different shades of grey in addition to black and white (see Fig. 3.1). Putting this in context, although the human eye can detect up to 190 different grey tones under the most perfect conditions, in normal circumstances only 50 different grey tones can be seen, so the number of different grey tones processed by a scanner is an important point to consider. As a scanner is only one part in the chain of what might be called a 'digitized image processing system', grey tones cannot be considered in isolation. The printer or imagesetter which we will typically use to print a scanned image works in a very simple manner, dot by dot, on the basis of the black and white values of the subject being printed.

Figure 3.1 Halftone scan with 16 grey steps

TECHNIQUE OF BITMAPPING

Shades of grey can be simulated when we work with black and white images. This simulation is done with a predefined arrangement of black and white points (Fig. 3.2) called a screen. Figure 3.2 shows a screen formed by 4 × 4 cells. Each of the 16 cells has more or less black and white in relation to the different grey shades to be simulated. Each grey shade is represented by a specific black and white pattern.

Working with screen cells in this way is called bitmapping, as each grey shade pixel is represented by a square cell with n × n dots. The single black and white dots are distributed differently and the density of such a screen cell depends on the number of the black dots. The potential number of different grey values in one screen cell are calculated with this formula:

$$n^2 + 1 = \text{Number of grey values in one screen cell}$$

The screen cell in Fig. 3.2 is formed by 4 × 4 single dots where n is set to 4. Therefore it is possible to represent 17 different grey shades.

Figure 3.3 shows an enlarged screen dump of a scan which is shown on the monitor using the bitmapping technique: all single screen pixels may only have the values black or white. If we consider the technique where screen cells represent different grey shades then we can see that a 2 × 2 screen cell enables us to use only five different grey shades whereas we can handle 65 different shades of grey with an 8 × 8 screen cell. As the number of greys is increased, the printing resolution is reduced. We will discuss this point later in some depth.

Figure 3.2 Example of a 4 × 4 screen cell representing 17 grey shades

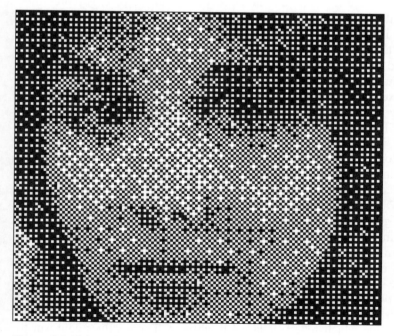

Figure. 3.3 Enlarged part of a bitmapped image on a monitor

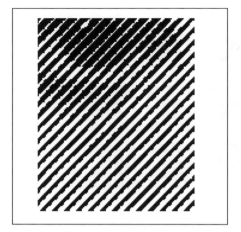

Figure 3.4 The line screen technique to represent grey shades

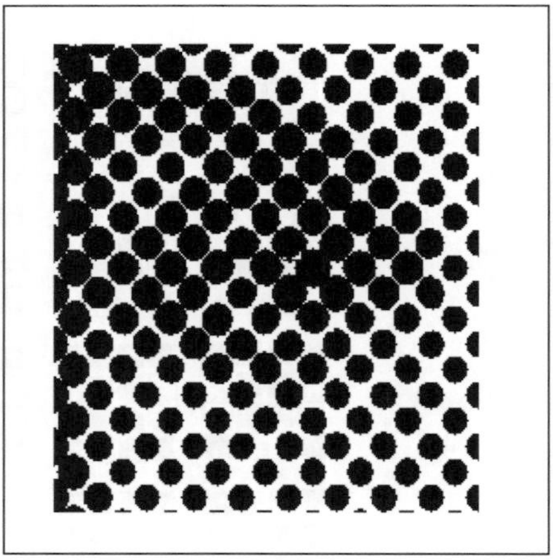

Figure 3.5 The pixel screen technique to represent grey shades

Although the bitmapping technique is normally used for simulating grey shades, there are other techniques available. The 'line screen technique' represents grey shades by means of diagonal stripes of varying widths (Fig. 3.4). Using the 'pixel screen technique' the density of an image pixel is represented by dots of different sizes (Fig. 3.5). This is used for photogravure printing in high quality magazines. For a given resolution wider screen cells with more image pixels are needed, compared with bitmapping.

The 'Floyd Steinberg technique' uses black and white single pixels whereby mistakes resulting from the simulated density are partly transferred to the adjoining pixels. This produces a black and white pattern that gives the impression of grey shades. The disadvantages of this technique are an unsettled pattern (see Fig. 3.6) and the relatively high amount of computing power needed.

With the 'ordered dithering technique' a black and white pattern is created by overlapping the grey value with a random number that goes above or below the black and white threshold. The impression is similar to that shown in Fig. 3.6. The last two techniques are only used for a rough representation of images with grey shades since lines or contours are suppressed or only partly shown. The number of grey shades that are used to represent a halftone image is also known as the 'depth of graphics'. In each screen cell, 4, 6 or 8 bits of information are used to store 16, 64 or 256 grey shades. The more shades of grey that are stored, the larger the amount of memory that is needed to store the image. Coloured originals are treated like halftone originals by a black and white scanner. During scanning each colour value in the original is converted to a digital value which reflects relative brightness. These values conform to different grey shades (see Fig. 3.7).

Figure 3.6 The Floyd Steinberg technique to represent grey shades

Figure 3.7 Coloured original for which colour values were converted to grey values

RESOLUTION AND GREY SHADES

The use of screen cells that contain a number of pixels reduces the resolution of the output device in proportion to the width of the cell. For instance, when a screen cell of 4 × 4 pixels is used, the resolution of a 300 dpi laser printer is reduced to 75 dpi, or a quarter of its maximum value. To expand upon this, a typical laser printer works with a resolution of 300 dpi or 12 lines of dots for each millimetre of printed copy. A screen cell with 8 × 8 pixels is needed for the representation of 64 different grey shades, and this reduces the resolution to 1/8 of the normal value, which is 37.5 dpi or 1.5 lines of dots per millimetre. This is a very poor resolution and we have to decide whether to print more grey shades at low resolution or less grey shades at high resolution. If we choose a higher resolution of 100 dpi (4 lines per millimetre) the following formula determines how many grey shades can be reproduced:

$$(\text{Maximum resolution / Effective resolution})^2 + 1$$

$$\text{Example: } (300/100)^2 + 1$$

The result shows that only 10 grey shades can be represented and this is unsatisfactory. Using 16 grey shades, where the resolution is only 3 lines per millimetre or 75 dpi, seems to be a reasonable compromise for the output of halftone images using a 300 dpi laser printer. Normal newspaper quality halftones have a resolution of 4 lines per millimetre. The resolution of monitors is lower than the resolution of laser printers and if it is necessary to show grey shades on a monitor, it is normally only possible if we use a part of a scan with less than the true number of greys.

HINTS FOR PRACTICAL USE

Halftone originals for outputting to a 300 dpi laser printer can safely be scanned at a resolution of 75–100 dpi with 16 or 64 grey shades. When the output is to be produced by a 1270 dpi imagesetter, we should use a scan resolution of 150 dpi with 64 or 256 shades of grey. As long as we expect to work with a laser printer with a resolution not exceeding 300 dpi, a scanner with 64 grey shades can be acceptable. Even this number of grey shades is only truly effective with high resolution imagesetters as they can output at a resolution up to 2540 dpi. We may have no choice in the matter as most scanners these days can scan at up to 256 shades of grey.

Now for some practical scanning examples. For the first example we scanned a halftone original with the HP ScanJet IIp with 16 and then with 256 grey shades. The scan resolution in each case was 75 dpi. Both unscreened grey scans were saved as TIFF files and they were imported into Aldus

PageMaker and printed without modification, using an Apple LaserWriter II NTX PostScript laser printer. Figures 3.8 and 3.9 show the results for the different number of grey shades.

Both examples show that the printer imposes its own limitations on scanned images. Scanners like the HP ScanJet IIp which can scan at resolutions between 16 and 256 grey shades should be used at the higher number. When a scanner is to be used exclusively with a 300 dpi laser printer, then one which can handle 64 shades of grey will be sufficient. Many scanners nowadays, however, have a minimum specification which enables scanning at resolutions of up to 256 greys, this makes the selection of a scanner easier than it has been in the past, and there may be no technical or financial advantage in seeking a lower specification.

We have talked so far about scanning images at their original size. Where an image, once scanned, is to be reduced in size, then a proportionally lower scan resolution can be used. Where the scan is to be enlarged, then a higher scan resolution needs to be used (see Figs. 3.10 to 3.12).

A common mistake when scanning halftone originals is to believe it is necessary to use the maximum resolution of the scanner. This highest resolution (i.e. 300 dpi) should only be used for line art and other originals without values for grey or brightness.

Figure 3.8 Halftone scan with 16 grey shades, scan resolution 75 dpi

Figure 3.9 Halftone scan with 256 grey shades, scan resolution 75 dpi

Figure 3.10a 16 grey shades,
75 dpi scan, output 1:1

Figure 3.10b 16 grey shades,
75 dpi scan, output 150%

Figure 3.10c 16 grey shades,
75 dpi scan, output 50%

Figure 3.11a 16 grey shades,
150 dpi scan, output 1:1

Figure 3.11b 16 grey shades,
150 dpi scan, output 150%

Figure 3.12b 16 grey shades,
37.5 dpi scan, output 50%

Figure 3.12a 16 grey shades,
37.5 dpi scan, output 1:1

FORMULAE FOR SCAN AND PRINTING RESOLUTION

There are two important formulae that help us to decide the right scan
resolution for halftone originals for a given printer resolution.

Formula 1

$$(\text{Size of screen cell})^2 + 1 = \text{Number of grey shades}$$

$$\text{Example: } 8^2 + 1 = 65$$

With a 8×8 screen cell it is possible to represent 65 grey shades.

Formula 2

$$\text{Printer resolution / Size of screen cell} = \text{Actual printer resolution}$$

$$\text{Example: } 300 \text{ dpi} / 4 = 75 \text{ dpi}$$

With a printer resolution of 300 dpi and with a 4×4 screen cell (for 17 grey
shades) the actual printer resolution will be 75 dpi (or 75 lpi = lines per inch)
or 3 lines per millimetre.

RESOLUTION AND LINE ART

The expression 'line art' describes graphics with a depth of one bit. Line art is usually printed at the original scan resolution since a line art object scanned at 100 dpi, for instance, is not improved when printed at 300 dpi. When scanning line art or contour drawings (see Fig. 3.13), or scanning halftone originals using a specific screen pattern (see Fig. 3.14) and using a printer of 300 dpi or less, we should choose a scan resolution which is the same as that of the printer. If we do not, then we may introduce distortion when the image is printed. Even if the program used to produce the output prints the scan at the original size, there may be annoying moiré patterns in the output.

If we want to scan line art or contour drawings to print at a resolution greater than 300 dpi, or if we want to send them to an imagesetter, we should use the highest resolution available on the scanner. When the scan program can scale a scan we should make the scan at original size for the best image quality. Even if the highest resolution for the scanner is limited to 300 dpi it is possible to get a higher resolution for the printed image. To do this the scan should be proportionally reduced in the printing program (for instance, in a desktop publishing program). With a reduction of 50 per cent, and sending the output to an imagesetter at 1270 dpi resolution, for instance, we will get an actual resolution of 600 dpi for the printed scan. If we reduce the image size to 25 per cent, then the printing resolution on the same imagesetter will go up further to 1200 dpi. As long as we are working with line art and contour drawings there is no danger of creating moiré patterns.

Scanning and screens

With commonly used monochrome output devices (matrix and laser printers as well as image- or typesetters) it is only possible to print single black dots on a white background, or vice versa. It is impossible to represent a single grey shade by a single printing dot. If we imagine the total printable area as a collection of many single dots, these dots can only have the values black or white. Therefore it is essential to screen an original containing grey shades, whereby each grey shade is associated with a bigger or smaller screen dot. In this way the human eye interprets grey values which are not really there. Screened images in newspapers are a common example of this (see Fig. 3.15).

To understand printing screens it is important to understand that a *screen* dot is made up of a number of *printing* dots. This is how it is possible to associate each screen point with different values corresponding to the number of printing dots. The different printing dots within a screen dot are always black. In relation to the number of the representable grey values a screen dot is composed of a larger or smaller number of printing dots. In the same way that printing resolution is described in dots per inch (dpi), screen resolution is described as lines per inch (lpi). This measurement describes the number of screen dots printable within an inch.

Figure 3.13a Line art, 100 dpi scan, 300 dpi print

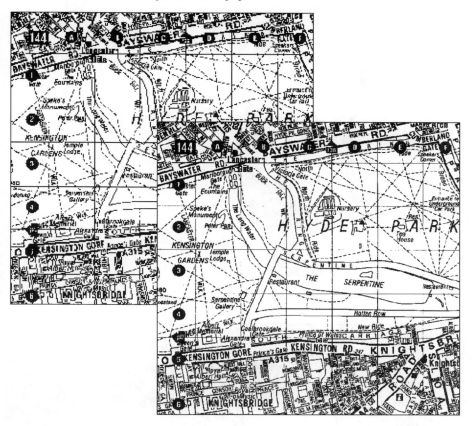

Figure 3.13b Line art, 200 dpi scan, 300 dpi print

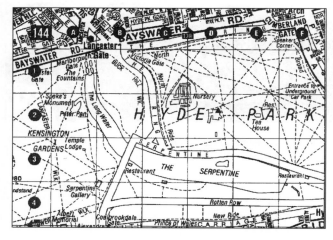

Figure 3.13c Line art, 300 dpi scan, 300 dpi print

Figure 3.14 Scan with screen pattern, 300 dpi scan and print

Figure 3.15 Enlarged section of a screened image in a newspaper

We have mentioned the different grey values which can be represented in a matrix with n × n image dots in connection with resolution; this is called a screen cell. This matrix is substituted during screening by another one with printing points for black and white; this is called a printing cell. The size of this matrix determines the number of grey shades which may be represented. Special algorithms are needed to convert screen cell information into a printing cell. For the conversion a reference screen is often needed which has the same number of rows and columns as the screen cell, the size of the reference screen is then identical to the size of the printing dot matrix. While screening the grey value, each image point within a screen cell is compared with the corresponding value in the reference screen and then converted to a printable or non-printable point within the printing cell. For this comparison different methods are available (for instance, threshold value, mean value or calculated value).

Selecting the printing screen

The selection of the correct printing screen depends on the resolution of the output device, the number of grey shades and the detail precision of the output. The last of these depends on the size of the halftone screen: the smaller the screen the more image details can be recognized. This is because of the fact that with smaller image point matrices it is possible to create smaller printing screens, which increases the resolution and the quality of reproduction. The disadvantage of small screen cells is the ability to represent only a limited number of grey shades. The bigger the halftone screen the more grey shades we can have. The relationship between screen width and resolution is calculated by the following formula:

Screen width = Printer resolution / Number of lines in the screen cell

The screen width is directly related to the resolution of the output device. Table 3.1 shows the screen widths with different printing resolutions and screen cells. The bold printed screen widths are ideal values for the appropriate resolution of the output devices. It is possible to use other values, but screen widths with more than 150 lpi cannot be supported even by output media like photographic papers or films: dark areas become too dense and light ones are broken off. Furthermore, in normal circumstances the human eye can recognize only 50 different grey shades which is why even an imagesetter need not output more than 64 grey shades. A laser printer with a resolution of 300 dpi can represent around 35 grey shades in good quality— although that is insufficient for professional usage.

Table 3.1 Screen width and print resolution

Screen cell	Grey shades	Screen width (lpi) with a resolution of		
		300 dpi	**1250 dpi**	**2500 dpi**
4×4	17	75	317.5	635
6×6	37	**50**	211.6	423.3
8×8	65	37.5	**158.7**	317.5
16×16	257	18.7	79.4	**158.7**

Table 3.2 Screen widths for printed material

Normal screen width for:	Screen width per inch (lpi)	Screen width per cm
Newspapers	65–80	25–30
Newsletters, books with low printing quality	80–130	36–48
Magazines, brochures, books with high printing quality	130–150	54–80

Screen widths are usually given in lines per inch. This value must be divided by 2.54 to convert it to centimetres, or by 25.4 to convert it to millimetres. Table 3.2 gives some idea of screen widths used for different types of printed material. It is important not to confuse printing resolution, measured in dots per inch, with screen width, measured in lines per inch. The screen width used for an output is determined by the number of shades of grey the printer is to produce or, to put it another way, by how many screen cells the printer needs to represent the different grey shades. The number of printable grey shades is calculated with the following formula:

$$\text{Number of grey shades} = \text{Printing resolution (dpi)}^2 / \text{Screen width (lpi)}^2$$

When, for example, an image file which is a halftone image is to be printed using a laser printer with a resolution of 300 dpi with a screen width of 50 lines per inch, the image will be screened in such a way as to represent 36 grey values. The bigger the value chosen for the screen width the denser are the lines and the fewer grey values can be represented. If the printing resolution is increased with the screen width, we will get more representable grey values. For example, if we want to output an image file using an imagesetter with a resolution of 1270 dpi, we should choose a screen width of 90 to 150 lines per inch.

Screen angle

Another important point is the selection of the correct screen angle to determine the run or the order of the single screen dots. The standard is an angle of 45 degrees because that value is nearest to the way our eyes see (see Fig. 3.16).

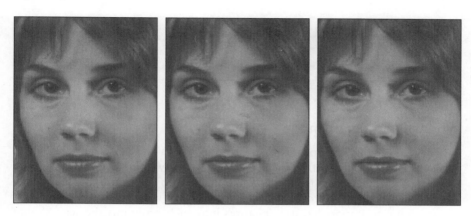

Figure 3.16 Different screen angles (0, 45 and 90 degrees)

Screen types

Besides the normal dot screen with an angle of 45 degrees it is possible to use different line screens and random screens for a printout. These screens are often used for special graphic effects when combined with the extreme screen widths of the normal screens (see Figs. 3.17 to 3.19).

Figure 3.17 Line screen for graphic effects (PageMaker)

Figure 3.18 Dot screen, screen width 15 lpi

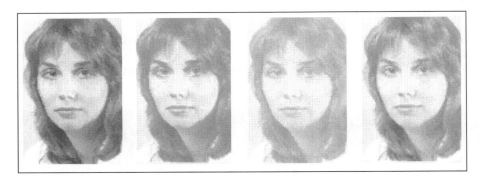

Figure 3.19 Random screens

When to screen an image

When we use a PostScript device to output a scan we can either screen the image before printing, within the scan program, or within the program we are using to manipulate and to print the scan. Desktop publishing programs like Aldus PageMaker or Xerox Ventura Publisher can work in this latter way. If we are going to import halftone pictures into such programs the scan should be produced and saved only with the necessary grey information and without a screen of any sort. The programs have facilities for manipulating the imported image file and for changing the image size. If we try to use an image file which has already been screened, then the screen would be destroyed during any manipulation and the resultant output would be rendered unusable.

SCAN RESOLUTION AND PRINTING

So, what resolution should we use to scan halftone originals to obtain the best results in printing? The following simple formula may help:

$$\text{Scan resolution} = \text{Screen width} \times 1.5$$

The screen width for this formula depends on the resolution of the output device:

$$300 \text{ dpi} = \text{around } 50 \text{ lpi}$$
$$1200 \text{ dpi} = \text{around } 140 \text{ lpi}$$

When we multiply the screen width by 1.5 we get the scan resolution we should use. Therefore, to output using a 300 dpi laser printer, we can scan at a resolution of 75 dpi. Although this seems a very low resolution—and we can only use it for halftone images—we can scan with 64 grey shades for an optimum output. If we are printing using an imagesetter with a resolution of 1270 dpi then a scan with 200 dpi and 64 grey shades is needed. Higher resolutions with halftone images do not always result in a better output; more memory is needed and the processing speed is reduced. With line art other rules apply: scan resolution should always be as close as possible to the resolution of the printer.

SUMMARY

In this chapter we have learned about the relationship between grey shades, scan and printer resolution. We have found the best resolutions for scanning line art and halftone images in order to produce the most pleasing outputs. We have appreciated that print screens are important, and that they are fixed during the scanning process or within a program used for further processing or manipulation of a scan. We have seen some formulae which help us to make sensible decisions when planning to scan.

Chapter 4
Originals for Scanning

When we consider buying a scanner, we need to ask ourselves some questions. What are we going to use the scanner for? What sort of facilities will we have with the one we choose? There are basically two kinds of subject that we want to scan. The first includes pictures or drawings in black and white and which feature no greys in-between. The second kind includes halftone pictures or photographs, containing tones from white through many shades of grey to black. Colour halftones can be considered together with monochrome ones, as the scanning process converts different colours to equivalent shades of grey.

LINE ART

Figure 4.1 shows a typical line art subject scanned using a hand-held scanner. Before we go into further detail, we had better explain about the pictures we have used in this book. We have given technical details with many of them, so it is clear which scanner was used, together with the resolution, the file format and the file size. All scans were printed using a PostScript laser printer with a resolution of 300 dpi. This small cartoon was scribbled with a blue ballpoint pen, so we had to set the brightness control on the ScanMan 256 darker than normal. The result was a scan with continuous black lines.

Figure 4.2 also shows line art, but with more graphic detail. The file size is four times larger than for fig. 4.1 though the original was the same size and had the same scanning resolution.

Figure 4.3—part of a circuit plan—shows a further increase in the amount of information in a line art subject. A hand-held scanner is quite capable of scanning all the fine lines, but a steady hand is needed to make a really satisfactory scan: this particular one was not very successful. In Chapter 5 we will give some tips to help us scan steadily with a hand-held scanner. This circuit plan includes details which are coloured red, but the LEDs of the scanner were unable to see them. We scanned the same circuit plan again, this time using a flatbed scanner (see Fig. 4.4), the HP ScanJet IIp, and it is interesting to see that it can detect the colour red much better than the ScanMan 256. This is because the LEDs of the HP ScanJet IIp have a higher red sensitivity.

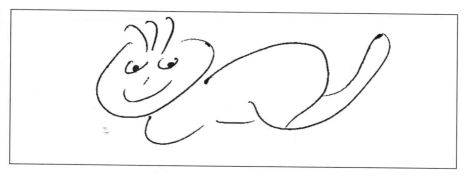

Figure 4.1 Line art subject

Scanning Information
Original: Line art, blue ballpoint pen
Scanner: Logitech ScanMan 256
Resolution: 200 dpi
Format: TIFF
File Size: 7458 bytes

Figure 4.2 More detailed line art subject

Scanning Information
Original: Line art, black and white
Size: 90 × 50 mm
Scanner: Logitech ScanMan 256
Resolution: 200 dpi
Format: TIFF
File Size: 30 975 bytes

Figure 4.5 shows a halftone scanned in line art mode. Here we get only black and white, as all grey tones are suppressed. Sometimes we need to make this kind of graphical conversion of a halftone or coloured original.

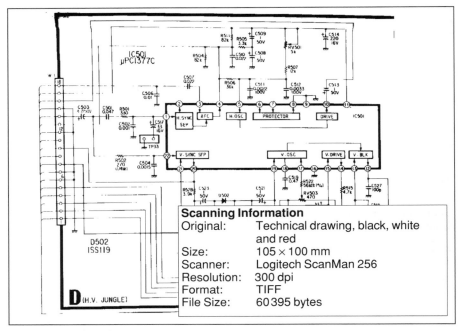

Figure 4.3 Circuit drawing

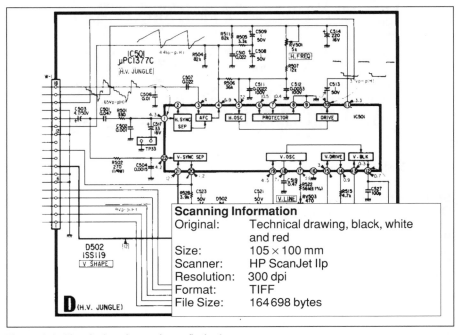

Figure 4.4 Circuit drawing, using a flatbed scanner

Scanning Information
Original: Photograph
Scanner: HP ScanJet IIp
Resolution: 300 dpi, line art mode
Format: TIFF
File Size: 127326 bytes

Figure 4.5 Black and white scan of a monochrome photograph

HALFTONE ORIGINALS

Scanning photographs is not in itself any more difficult than scanning line art, but we need to make more decisions to get a scan of the desired quality; we gave basic information on this aspect of scanning in Chapter 3. Whereas line art needs only black and white for faithful recording, a photograph or halftone consists of a wide range of tones representing relative brightness ranging from white to black. A subject which is just black and white has very little room for variation except in the relative proportions of those two colours, but each halftone subject can be very different in the relative number of grey values. Thus there are potentially very many more adjustments which may have to be made on the scanner or within the scanning program to achieve the result we want. We can use these adjustments to modify a halftone drastically to get a special effect.

CONTRAST AND BRIGHTNESS

The ideal black and white photograph contains balanced contrast with an even brightness. As this kind of ideal original is seldom available, we must make corrections for brightness and contrast during the scanning process, while manipulating the scan with appropriate software or, at the latest, before printing. At first, it may take a while to identify the best settings for a particular scan, but as we become more experienced we also gain a feeling for the adjustments that it will be necessary to make to different originals. To explain the relationship between the contrast and the brightness of an original, the scanned version, and the printout of the image, we will use a practical example.

THE IDEAL HALFTONE ORIGINAL

To show an ideal halftone original we have produced a series of prints of different brightness and contrast from a single black and white negative. We scanned these originals in grey scale mode, with 256 grey steps, leaving brightness and contrast set at default within the scanning program.The series of pictures in Figs. 4.6 to 4.8 shows that the ideal halftone original contains normal brightness and contrast (see Fig. 4.7b). This is not always appropriate as contrast also affects the mood of the photograph. A subject, for instance, shot in the 'hard' southern sun, with extremely strong light and shade often demands a soft contrast for an ideal result, whereas a subject shot on a foggy day demands hard contrast. The brightness of a picture is a different matter, and should always be normal, or well balanced.

ADJUSTMENTS BEFORE SCANNING

In black and white photography enlargements of negatives with hard or soft contrast are made by the selection of an appropriate photographic paper. The brightness of the enlargement is determined by the exposure time: dense negatives need a longer, and soft (bright) ones, a shorter exposure time. The development time is always constant. The same adjustments can be made before and after scanning. This chapter shows how we can make corrections to originals before we begin to scan.

Suppose we want to scan a halftone original with the ScanMan 256 hand-held scanner, where the adjustments are made directly on the scanner. We adjust for the brightness of the scanned original using the contrast control, which influences both the brightness and the contrast. For a stronger contrast we turn the control to the light mark, while for a weaker contrast we move it to the dark mark. We need to bear in mind that a light picture will give us more detail than a dark one. When a large picture is to be reduced after scanning it is best to scan with a light setting: as an image is reduced, so the picture gets darker.

Figure 4.6a Bright with softer contrast

Figure 4.6b Bright with normal contrast

Figure 4.6c Bright with harder contrast

Figure 4.7a Normal brightness with softer contrast

Figure 4.7b Normal brightness with normal contrast

Figure 4.7c Normal brightness with harder contrast

Scanning Information

Original: Photograph
Scanner: HP ScanJet IIp
Resolution: 75 dpi, 256 grey scales
Format: TIFF
File Size: 17981 bytes

Figure 4.8a Dark with softer contrast

Figure 4.8b Dark with normal contrast

Figure 4.8c Dark with harder contrast

In the better scanning programs we can control brightness and/or contrast from within the program itself, so we can adjust originals that are too dark or too light before we start scanning. The final result of these adjustments can be seen on the monitor, though it is very difficult to gain an accurate impression of how a scan will look from this, as a number of factors can influence the appearance of the image as it finally appears:

- The brightness and contrast of the scanner (controlled by hardware and/or software)
- The brightness and contrast of the monitor (controls)
- The brightness and contrast in the program that manipulates the scan at a later time
- The brightness and contrast in the program that will print the scan.

Constant scan results can only be guaranteed by making a series of tests, and noting the different settings. The first thing to set is the monitor, which should be adjusted to a contrast and brightness that is comfortable for our eyes. The settings of the controls should be marked. Next we can produce our first test series by choosing different settings for brightness and contrast in the scan program. Once we have done this, we can print each scan using the scanning program, or the program that will be used to print most scans, using default settings for brightness and contrast. After the first test series we will have developed a feeling for the ideal original and the sorts of corrections we will need to make, before or after the scanning process.

SUMMARY

In this chapter we have learned about the importance of the original we want to scan. We know that differences in the brightness and contrast of the original can be adjusted at the scanner or in the scanning program. The following chapters will show in detail the many adjustments we can make to a scanned picture.

Chapter 5
Hand-held Scanner: ScanMan 256

Logitech has been calculated to have more than half of the hand-held scanner market for grey scale scanners, and Logitech's ScanMan 256 is typical of the hand-held scanners on the market today (Fig. 5.1). After we have installed the scanner (see Chapter 2), it can be used with excellent speed and precision, and we will give some tips on how to improve the effectiveness of the scanning process in this chapter. The scanning window is 105 millimetres wide which makes it possible for us to scan originals up to 36.75 centimetres long and 10 centimetres wide (14 × 4.10 inches). This may seem a bit restrictive when we know that we can scan originals up to A4 size on a flatbed scanner, but the practicalities of scanning make this less important than it may seem. For instance, when we scan even quite small halftone originals at a high resolution (300 dpi) then we discover right away the restrictions imposed on us by our system RAM and the capacity of the fixed disk. Originals of standard colour print size (10 × 13 centimetres) pose no problems and are ideal for scanning with a hand-held scanner. We have also found that most line art is of a length and width suitable for scanning using such a scanner.

The ScanMan 256 package contains the scanner itself, the interface board (there are different versions for PC, XT, AT and PS/2) and the ScanMan and FotoTouch programs. Both of these programs are well documented. Earlier models in the hand-held scanner ranges would often be supplied with software that was based purely on DOS, and Logitech's ScanMan 32 can still be purchased in a DOS variant. Recently, hand-held scanners have followed the practice of earlier, more expensive, desktop scanners and have utilized the Microsoft Windows environment with great effect. This has made the use of a mouse mandatory if we want to gain full use from our scanning package.

The scanning resolution for the ScanMan 256 is selected by moving a switch at the right of the scanner to either 1, 2, 3 or 4 (see below for details). A brightness control dial adjusts the contrast between the light and dark parts of an image; adjusting the thumb wheel towards the light rectangle gives a lighter image, and adjusting it towards the dark rectangle will give a darker image. Another switch enables us to select between three different grey scale settings for halftone originals, and a line art setting.

The scanning window lets us see what we are scanning when the scanner light illuminates and the scanner is activated by the scanning program. Small

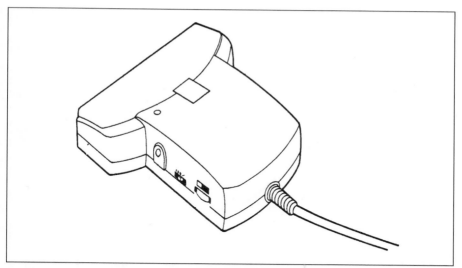

Figure 5.1 Logitech ScanMan 256 Hand-held scanner

Scanning Information
Original: Line art, black and white
Scanner: ScanMan 256
Resolution: 200 dpi line art
File format: PCX
File size: 13718 bytes

guides at the front and sides of the scanning window help us position the scanner over the original.

A small indicator light on the scanner itself helps us to control the scanning speed. The light comes on when the scanner is ready and it changes colour according to the scanning speed. If the scanning speed is correct, then the light is green; it changes to amber as the scanner nears the limit for the safe reading of the image, and turns to red when the scan is being made too fast. This information is echoed on the screen in the scanning window so that we need not have to take our eyes off the screen. Most of the time we have our eyes on the scanner itself, to ensure that it is tracking smoothly.

HOW TO HOLD THE SCANMAN 256

The scanner may be used in the left or the right hand. In the right hand the scanner is held so that the thumb is used to press the scan button (see Fig. 5.2). In the left hand it is held so that the Scan button is pressed with the middle finger. The arrow in Fig. 5.2 shows the direction in which the scanner is being moved over the original.

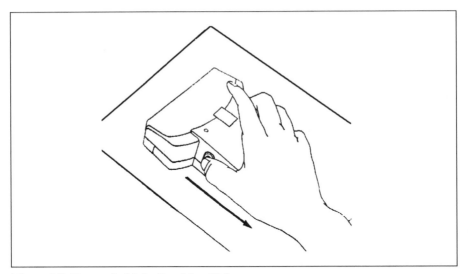

Figure 5.2 How to hold the ScanMan 256

Scanning Information
Original: Line art, black and white
Scanner: ScanMan 256
Resolution: 200 dpi line art
File format: PCX
File size: 21 807 bytes

The program-controlled start or triggering of the scanning process turns the scanner on and the scanning window in the ScanMan 256 is illuminated. At this point the scanner needs to be positioned on the original so that the scanning window is just above the image. We use the guides at the front and sides of the scanning window to position the scanner, but even with these aids it is quite difficult to move or roll the scanner precisely and in the right direction over the original. A straight scanning line can be made easier by using a ruler or the edge of a book to guide the scanner. This helps us avoid a slanted image with 'jaggies' or jagged lines, but for regular scanning this *ad hoc* arrangement is not satisfactory. We made up a simple 'scanning guide' as shown in Fig. 5.3, and found it very effective. Such guides may also be purchased. A scanning guide permits us to move the scanner precisely, and we can align and tape small originals to the work surface with great accuracy. We can then scan originals in landscape orientation more easily. One of the main problems with using a hand-held scanner is that of scanning small originals such as photographs, as the scanner itself can completely cover the original being scanned, causing movement and distorted results. Also, where images such as photographs need to be scanned close to the edge of the original, the leading edge of the hand scanner can catch on the original and refuse to go further. We found that this was not a problem when using the guide, or when we put the original inside a transparent sheet or folder.

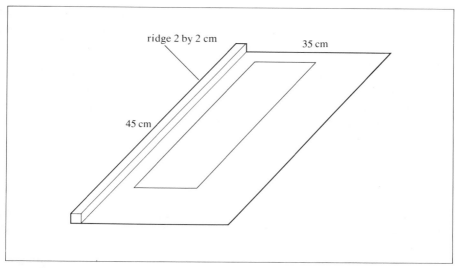

Figure 5.3 A scanning guide

TIPS FOR HAND-HELD SCANNING

Before we start scanning in earnest there are some tips that can simplify scanning with a hand-held scanner.

Scanning Originals

It is possible to scan just about every kind of original as long as it is flat and fits the maximum scanning width and length. Our choice of images can include cartoons, photographs, line art, logos, drawings, graphs, charts, diagrams, text, and so on. We can scan sections from flyers, magazines, books, newspapers, catalogues—just about anything that can be laid flat on a desktop or scanning desk. We can scan colour pictures or pictures with shades of grey, and we can scan black and white images such as text, line art or clip art. The original image should have reasonably good contrast. With such freedom to import images into our computer we need to ensure that we do not violate any copyrights that may exist on the originals. In general, we are safe if we are using the results of our scanning for purely personal purposes.

When an original is too big for the scan window, we can reduce it on a photocopying machine so it fits the scanning size of the ScanMan 256. For line art and pure black and white images this will result in no loss of quality. Too large an original can also be scanned as a series of strips which can be combined using an image processing program to produce something approaching the original size. The Logitech ScanMan 256 is supplied with such a program, and this is undoubtedly a major point in its favour.

Without such sophisticated means to combine the parts of an image it is almost impossible to avoid overlapping the different parts, making it necessary to perform further corrections and making the whole process difficult and time-consuming. Scanning the original in several passes on such scanners is only recommended when the original is divided in logical columns. We will delve a little further into this aspect of scanning later on in the chapter. Scanning an image with a glossy surface may result in an uncontrolled reflection of the light used by the scanner for the scanning process and this can generate splotches. To correct this, we adjust the brightness control dial on the scanner, or photocopy the original and then scan that.

Scanning Speed

It is necessary to use trial and error to determine the correct scanning or rolling speed. Generally with the ScanMan 256 we should scan at an even rate of 1 centimetre to 5 centimetres (1/2 inch to 2 inches) per second. For complex images such as halftone originals with many grey shades, or while scanning in high resolution (for instance 300 dpi) a slower speed is necessary. If the scan is made at too fast a speed, the scanner cannot receive all the incoming information and the image will be distorted or squeezed vertically. The scanner misses parts of the image, or hopelessly distorts it when we scan too fast, and we have to start again, scanning with a slower movement.

Scanning motion

It is very important that we hold the hand-held scanner steady whilst scanning. If we move the scanner in a jerky way, or pause during scanning, we will create jagged or wavy images. While this may be a useful capability when we are trying to achieve surreal or exaggerated results, it is clearly not a good thing when attempting precise reproduction. We must also avoid pressing the scanner too hard on the original. A precise, fluid movement at the correct speed guarantees the same quality results as a flatbed scanner.

Scanning resolution

Logitech recommends a standard resolution of 200 dpi. The required resolution must be selected on the scanner itself using the resolution switch on the right-hand side and this is recognized automatically by the scanning software. The different switch positions have the following meanings:

1 = 100 dpi   2 = 200 dpi   3 = 300 dpi   4 = 400 dpi

When we plan to output the scan using a laser printer we should select a resolution of 300 dpi. If we need scans for a presentation on a monitor the resolution selected can be closer to the resolution of the monitor (100 dpi).

Colour originals

Certain colours of an original may not be recognized by the scanner although this is minimized by the red light in the scanning window of the ScanMan 256. If colours are lost, we can rotate the thumb wheel on the scanner to get a harder (darker) or softer (lighter) contrast. Another solution which can help is to make a black and white photocopy and use that for scanning. We can use similar procedures for all other types of black and white scanner.

JUDGING THE SCAN

Up to a certain point the appearance and quality of a scanned image can be determined using the scanning and image processing software, and seeing the results on the monitor. The final proof will always be when the image is printed. The best way to get a feel for what we are doing is to scan an image with different settings for resolution, brightness and contrast. Save each scan, and write down the values that have been used in the process. Once the image has been printed, its appearance on screen may be compared with the final print and the settings that have been used. In this way it should be possible to judge the correctness of an image at the time of scanning from the representation of the image on the monitor. In such test series it is important not to adjust the brightness and contrast on the monitor itself during the different scanning steps. If we note all the adjustments to the monitor and within the scanning programs, then we will find it much simpler to produce a satisfactory result.

SCANNING WITH SCANMAN SCANNING SOFTWARE

In the following sections we will show how to make a scan using ScanMan scanning software. We will assume that the ScanMan 256 hardware has already been installed correctly (Chapter 2), its software has been loaded on to the disk, and there is a Windows icon we can use to launch it (Fig. 5.4). ScanMan scanning software cannot be run in isolation, it must be launched from within another application program, such as Word for Windows, or the FotoTouch program supplied with the scanner. As a first step, start up Windows. Using the mouse, double click on the Logitech FotoTouch icon to launch FotoTouch itself, then select *Acquire* from the FILE menu, and the *Logitech ScanMan* window appears (Fig. 5.5).

The *Logitech ScanMan* window

What we do next depends on the kind of image we want to scan. If it is a simple one that can be scanned in one pass from the top of the image to the

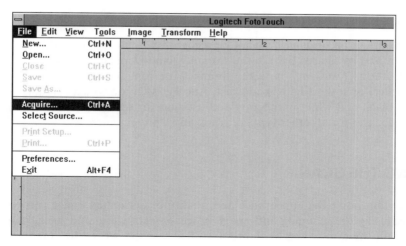

Figure 5.4 Acquiring the scanning window from FotoTouch

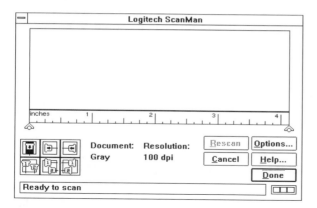

Figure 5.5 The scanning window from within FotoTouch

bottom, then we select the appropriate icon in the Scan Mode Panel (Fig. 5.6). We can also scan in a single pass from left to right, or from right to left. We can also scan in multiple passes from top to bottom, from left to right, or from right to left. The *ScanMan* window automatically shows us the resolution that is set on the scanner, and the mode—grey or black and white. These settings cannot be changed by software, only by means of the switches on the scanner.

There are a number of controls within the window area. We can, for instance, immediately see a ruler. This enables us to affect the width of the scanned image by cropping unwanted parts. This must be done empirically; that is, we should scan first, identify what needs to be cropped, then rescan as close to the previous path as possible. If we don't do it at this time, we can

Figure 5.6 Different directions for scanning

probably crop the image in the application in which we are going to use the scan. It is best, for reasons of image size on disk, to crop at scan time. As soon as the scanning window has appeared, the scanner lights up. If we take too long in starting to roll the scanner, the light goes out, and a dialogue box appears on screen to let us know this (Fig. 5.7). In this way, the software ensures that we don't leave the scanner activated for long periods, thus reducing its life.

Figure 5.7 Inactivity warning

Suppose we want to scan an image in one go: that is, in Logitech terms 'single-strip scanning'. By default the correct icon is usually selected in the scanning window (see Fig. 5.5). Now we position the scanner over the start of the image, press the scan button and release it, and move the hand scanner slowly and carefully over the original. As we are scanning the image begins to appear on the monitor (Fig. 5.8).

If we move the scanner too quickly over the original, the scanning speed indicator light will turn to red, and we will have to start again, as whatever we have tried to scan will probably be useless. Often, the scanner will have recorded nothing. If the scanner cuts out even at the right scanning speed, it is most likely that there is insufficient free memory for the image being scanned. Once a successful scan has been made, pressing the scan button again will turn off the scanner. We then need to click on the *Done* button in the scanning window, and the image will appear within the main *FotoTouch* window, ready for saving or manipulation (Fig. 5.10). To save us needing to stop the scan in this way, there is an option in the scanning software that will cause the scanner to halt the scan automatically when the scanner has stopped moving for a couple of seconds (Fig. 5.9).

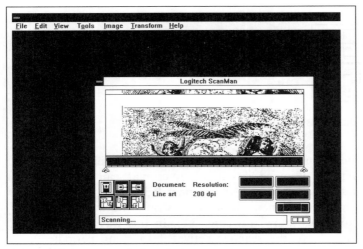

Figure 5.8 Scanning window during scanning

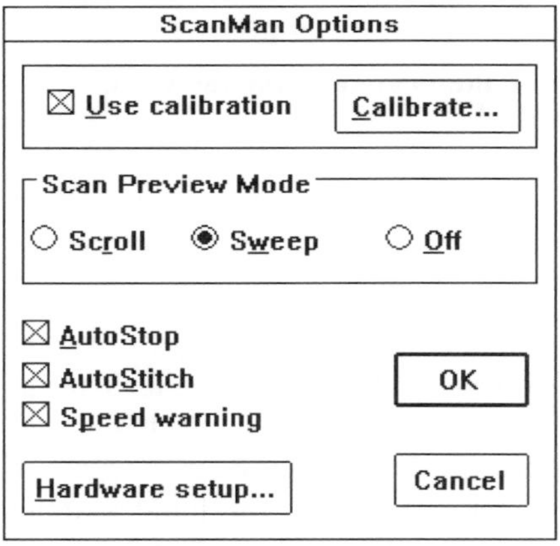

Figure 5.9 Option window to select auto stop

When the quality of the image we have scanned suits us then we can save it
for processing later. Such processing might be simply to print it, or it might be
to enhance it in some way. We will discuss this in later chapters. To save a
scan we open the FILE menu and choose the option *Save As*. By default,
FotoTouch saves a scan in the TIFF format (Tagged Image File Format), and
it will automatically append the extension .TIF to the file name. The default
TIFF file will be saved in compressed form, thus saving disk space. Other
options available include TIFF uncompressed, Windows Bitmap (BMP), PC
Paintbrush (PCX) and Encapsulated PostScript (EPS). The TIFF (compressed
and uncompressed), BMP, PCX and EPS file formats can be used without any
problems by desktop publishing programs such as Aldus PageMaker.

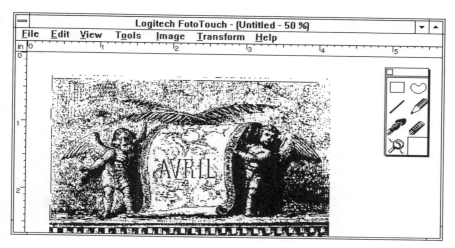

Figure 5.10 The completed scan, displayed in the scanning window

IMAGE SIZE AND MEMORY

The size of image it is possible to scan using the ScanMan 256 depends wholly on the amount of memory on our computer system. Each scanned image is initially stored in the main memory of the PC, and with Windows 3.1 we have a much greater potential for larger images than we had using DOS without Windows. However, every system has its limits in terms of memory. Overall, the maximum length a scan can take up is 22 inches, and this can lead to enormous files if we choose to make high resolution scans of this length. The final image size is also affected by:

- Defined scan width
- Document mode, grey scale or black and white
- Scanning resolution
- Format we select to save the image.

As an example, if we take the image we scanned earlier, and save it in each of the available formats we get the following results shown in Table 5.1:

Table 5.1

File format	File size (kb)
TIFF compressed	250 884
TIFF uncompressed	339 380
BMP	334 798
PCX	346 840
EPS	668 557

The image had been scanned at 64 grey levels, and 200 dpi.

Logitech provides some helpful information to enable us to calculate image size for images saved using uncompressed TIFF format. The file size in kilobytes will be:

$$\frac{\text{Image length} \times \text{Image width} \times \text{dpi} \times \text{Bits per pixel}}{8 \times 1024}$$

As a further guide, Logitech provide the information shown in Table 5.2 to enable a ScanMan user to estimate file sizes for uncompressed TIFF. These calculations are for an image 1 inch square. Clearly, we can make our calculations by working out the number of square inches in our image and multiplying that number by the appropriate figure.

Table 5.2

Resolution	Black and white	Grey scale	Colour
100 dpi	1.22Kb	10Kb	30Kb
200 dpi	5Kb	40Kb	120Kb
300 dpi	11Kb	90Kb	270Kb
400 dpi	19.5Kb	160Kb	480Kb

MANIPULATING THE PICTURE IN FOTOTOUCH

The ScanMan program in the ScanMan 256 package can only be used for scanning but FotoTouch contains some useful capabilities for picture manipulation that we will describe using the following example. FotoTouch is a fully-featured paint program that has additional support for the ScanMan 256 so that we can scan directly into FotoTouch from our desktop. We won't describe every feature of the program; we will only indicate the options for manipulating and refining a scanned image. FotoTouch brings the images directly into its work area on screen. We can zoom in on the image, or zoom back out using the VIEW menu, or by pressing '+' to zoom in, and '−' to zoom out.

The image we have used for demonstrating zooming is a black and white one. We get an equivalent and quite spectacular effect on screen when we are scanning and viewing a grey scale image. Unfortunately, our screen saving capabilities do not permit us to show this directly, only the final image when we export it to another application and print from that, or when we print directly from FotoTouch (Figs. 5.11 and 5.12). The resolution of the monitor and the size of the scan itself defines just how much of the scan will be shown on the screen without zooming. For instance, if our monitor has a resolution of 72 dpi, and we are scanning at a resolution of 200 dpi, then the monitor will

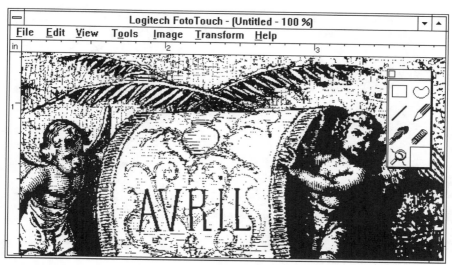

Figure 5.11 Image, after minor zoom

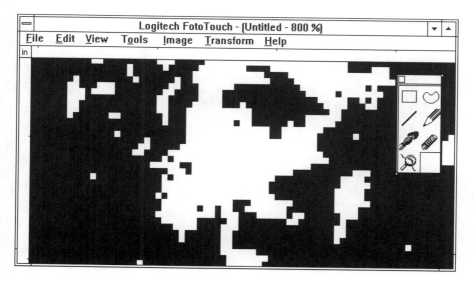

Figure 5.12 Image, at maximum extent of zoom

only be able to show part of the scanned image. We can move around the image at this size, using the scroll bars; we can move the image using the hand tool (🖐), or we can zoom out (🔍) to see more of the image.

FotoTouch tools include an eraser (✐) that we can use to remove unwanted parts of an image (Figs. 5.13 and 5.14). We can also use a paint-brush (🖌) and pencil (✏) for manipulating and refining the scan. For both

Figure 5.13 Revising a scan with the eraser tool

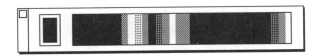

Figure 5.14 Original and edited scan

Figure 5.15 Paint palette

of these tools we can choose patterns from the paint palette in the main screen (Fig. 5.15). These tools may be used at any of the zoom ranges that this software permits. This can be at the pixel level in a black and white scan at maximum zoom. When we want to use the tools for very fine detail, zoomed close in, we hit a slight snag in that we can only use them on the part of the image that is visible in the main screen. If we want to paint or spray outside that part of the image, we have to scroll around the image using the scroll tool to bring other bits of the image into the main screen. The main disadvantage in this kind of manipulation of images was the need to interrupt our work, move to another part of the image and to continue to work while avoiding making obvious joins where blending took place.

The central picture in Fig. 5.16 shows a grey scale image that has been scanned at 256 grey scales and 100 dpi at a particular level of brightness. Next to it to the left is the same image after it has been adjusted with 80 more grey scales in FotoTouch. The picture to the right has been adjusted with 40 less grey scales in FotoTouch. The options in the IMAGE menu enable us to *Lighten*, *Darken*, *Smooth*, *Sharpen*, reverse the shades (*Negative*), evenly distribute the shades (*Equalize*) or posterize (*Threshold*) selected image areas in the main screen, or for the whole screen if no selection has been made (Fig. 5.17). To give us better control when we are changing the brightness and contrast of our scanned image, FotoTouch provides us with a set of sliders that show us, in real time, the effects of changes to the settings (Figs. 5.18a and b). If we want to transform an image or part of an image in some way— say, to rotate it, then we use the appropriate options in the TRANSFORM menu. The options include *Flip Horizontal*, *Flip Vertical*, *Rotate Left*, *Rotate Right* and *Deskew*. Combinations of the options may be used: Fig. 5.19 shows, for example, the same scan reversed, then rotated.

Figure 5.16 Image as scanned and with brightness adjusted -80 grey scales and +40

Figure 5.17 Image set to *Threshold*, softened, sharpened within FotoTouch

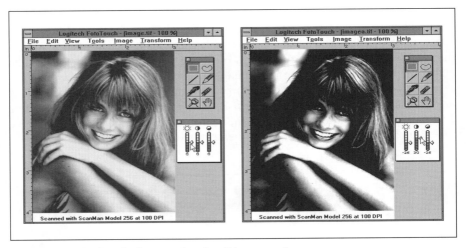

Figure 5.18 Altering the image using the slider controls

When we have a line art scan, a pixel by pixel modification is possible by zooming as far as possible, and using the pencil tool. Pixel colour can be set by clicking the mouse on the shade panel, and additional pixels can be set using the left mouse button. When we press the '–' key, we zoom out from the image. Pixel modification is also possible with halftone scans but this takes a lot of experience and time, and may not always be effective. The only way to learn what can and cannot be done in modifying a scanned image is to experiment with what we have described above. For safety's sake, always save the original scan first, as we can then return to it in the event of our making a

Figure 5.19 A reversed and then rotated scan

major mistake in the editing process. This is much quicker than having to scan the image a second time. We found the most useful features in FotoTouch to be those that enable us to adjust brightness and contrast. Though it is possible to edit at pixel level within FotoTouch, we prefer to use PC PaintBrush, supplied with Windows 3.1 for that particular task.

SCANNING IN MULTIPLE PASSES

The main limitation in using a hand-held scanner can be the relatively small scan width which is not enough for an A4 sized original. The ScanMan software can help us with wider images as it has an AutoStitch capability. The idea of this is that we first select multiple strip scanning from the scanner menu, and scan the first strip of our image. We then scan a second strip, allowing an overlap of between 0.5inch and 1.5inches with the first strip we scanned. The software then locates where the two strips join, and produces a single image. It appeared from a first glance at the documentation that AutoStitch only works with 256 grey scale scans, and that these need to have a resolution of 400 dpi to be successful. Whenever we tried to use these settings, we never had enough space to hold the image. However, when we reduced the scanning resolution to a more usual 200 dpi, the AutoStitch worked, but the resulting image was too poor to use (Figs. 5.20a and 5.20b). Subsequent editing, sampling the colours around the stitch marks, made the picture far more acceptable. Subsequent tests showed that AutoStitch also worked with line art scans, although we always needed to make a slight cosmetic change to the final results.

Figure 5.20a AutoStitched photograph

Figure 5.20b AutoStitched photograph, after editing

PRINTING WITH FOTOTOUCH

FotoTouch has a limited printing capability. To print an image, we open the FILE menu and choose the option *Print*. The appropriate printer will have been defined during the Windows set-up. When we click on *Print*, we see the initial PRINT dialogue box (Fig. 5.21). For a simple print, we can just click on *Print* in this window. Alternatively, we can change the position of the image on the page, by adjusting the margins. For more flexibility in printing, we can click on *Output Format*.

In the *Output Format* window shown in Fig. 5.22 we can select the number of greys we want to print, place a box around the image when we print it, and/or scale the image. If we wish to frame the picture, there are a variety of frame styles at our disposal (Fig. 5.23).

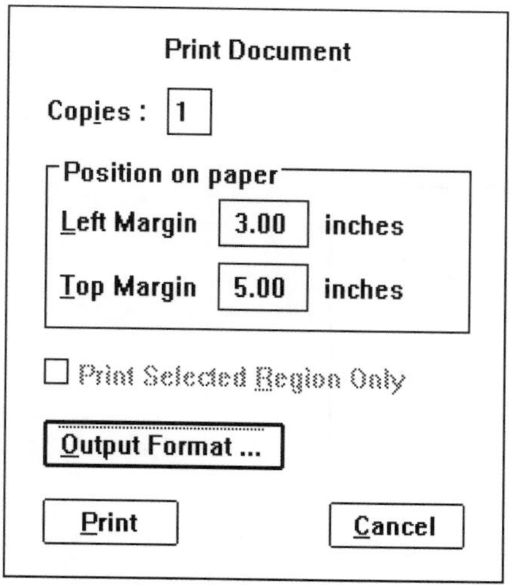

Print Document

Cop<u>i</u>es : `1`

┌─ Position on paper ─────────┐
<u>L</u>eft Margin `3.00` inches

<u>T</u>op Margin `5.00` inches
└──────────────────────────┘

☐ Print Selected Region Only

`Output Format ...`

`Print`   `Cancel`

Figure 5.21 Initial *Print* dialogue box

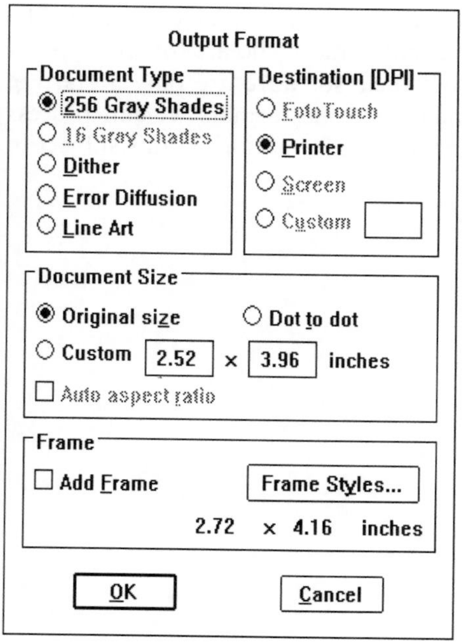

Output Format

┌─ Document Type ─┐   ┌─ Destination [DPI] ─┐
◉ 256 Gray Shades   ○ FotoTouch
○ 16 Gray Shades   ◉ Printer
○ Dither   ○ Screen
○ Error Diffusion   ○ Custom `___`
○ Line Art

┌─ Document Size ─────────────┐
◉ Original si<u>z</u>e   ○ Dot <u>t</u>o dot
○ Custom `2.52` × `3.96` inches
☐ Auto aspect ratio
└──────────────────────────┘

┌─ Frame ──────────────────┐
☐ Add <u>F</u>rame   `Frame Styles...`
          2.72 × 4.16 inches
└──────────────────────────┘

`OK`   `Cancel`

Figure 5.22 *Output Format* window

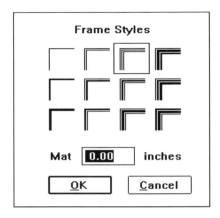

Figure 5.23 Different frame styles for printing

SUMMARY

In this chapter we have introduced the hand-held scanner and shown how we can use it, and also how we can manipulate the images we have scanned. The Logitech ScanMan 256 represents the current state of the art for this type of scanner. We have seen that scanners and their software are only a starting point. Because scans are stored in standardized file formats we can process them using other programs. There will be more information on this subject in later chapters within this book.

Chapter 6
Flatbed Scanner: HP ScanJet IIp

The Hewlett-Packard ScanJet IIp (Fig. 6.1) is a typical flatbed scanner and the ScanJet range of image scanners probably accounts for around 40 per cent of the flatbed market. Versions are available for the AT-bus systems, also known as ISA architecture systems, and for MicroChannel systems (MCA) such as IBM's PS/2 range, models 50 and above. We used an AT-bus IBM-compatible system. The ScanJet IIp is a high resolution 8-bit grey scale scanner able to recognize up to 256 grey scale levels, with numerous settings for contrast and intensity (brightness) in halftone, screening and line art modes. It can be used to scan full pages up to 216 × 297 millimetres. Line art, and screened originals can be scanned in A4 size at 300 dpi within 10 seconds. The optical resolution will be 300 dpi while the output resolution can be set in dpi steps from 12 and 1500 dpi. This can be confusing, and it should be appreciated that originals will always be scanned at a maximum of 300 dpi. The higher resolutions are calculated by interpolation; that is, additional points are placed between two points which have been scanned.

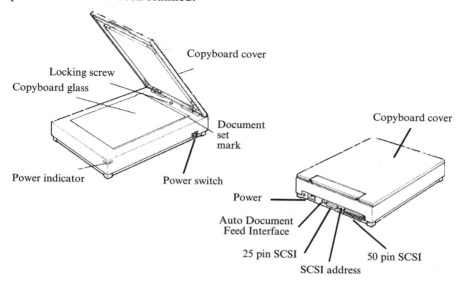

Figure 6.1 The Hewlett-Packard ScanJet IIp flatbed scanner

An automatic document feeder option is available, to enable up to 20 pages to be scanned, one after the other. This feeder is essential if we intend to use the ScanJet IIp for Optical Character Recognition (OCR), for which special software is needed.

The ScanJet IIp package consists of the scanner itself, a good user hand-book, the DeskScan scanning software and PhotoFinish image processing software as Microsoft Windows applications. The software includes the device drivers for various ScanJet models, the Scantest program, an AT or MCA compatible interface card and an interface cable for connecting the scanner with the interface card (Fig. 6.2).

The ScanJet IIp device driver is a program that enables communication between the scanner and the scanning programs. Scantest is a Windows program to test the scanner, and it tells us whether the installation of the scanner and the interface card has been made correctly. This program also checks the scanning precision. DeskScan is used for scanning and editing (to a degree) images. PhotoFinish provides far greater image manipulation capabilities. All of these programs run in the Windows 3.1 environment, and not as DOS applications. A full version of Windows 3.1 is needed to be able to run the Scantest and DeskScan applications. Earlier models of the ScanJet could be supplied with runtime versions of the earlier Windows 2. This is no longer possible as applications have grown to utilize the capabilities of the much-enhanced Windows 3.1. This should be no problem, as the majority of new personal computer systems these days are supplied with Windows 3.1.

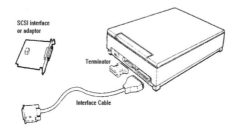

Figure 6.2 Hardware of the ScanJet IIp package

INSTALLATION

The installation of the HP ScanJet IIp is performed as a sequence of different steps:

- Installing the DeskScan software on the computer system
- Setting up the interface card
- Installing the scanner interface card

- Connecting the scanner to the interface card
- Testing that the scanner has been correctly installed, and works with the computer.

The following description of scanner installation is for an AT system. Installation for MCA and EISA-bus systems differs a little but is described in detail in the ScanJet IIp user handbook.

Step 1

Boot your PC if you have not already done so, and start up Windows 3.1. In the Program Manager, select FILE and click on *Run*. Insert the first DeskScan diskette into drive A. Both formats of floppy diskette are supplied, so this should cause no problem. We can, if necessary, use drive B if, for instance, we corrupt one set of diskettes, and we need to use the other set. On the command line, type A:install and either press ENTER, or click on OK. The installation will then proceed and we will be prompted each time a new diskette needs to be loaded. At the end of the installation, we will be asked if we want our CONFIG.SYS, AUTOEXEC.BAT and WIN.INI files to be changed to reflect the installation. Some people don't like to have installation programs changing their key system files, and if we decide to say NO to such changes, we will be given details of how to make the changes ourselves. If we say YES, then the original files will be saved with an extension of .OLD, so we can always get back to the original state. We can now shut down Windows, and revert to DOS.

Step 2

The next step is to set up the interface card. This card is very small and the principal feature on it is a block of four dip switches that are used to select the address to be used. When an IBM XT, AT or compatible computer—or an EISA system—is to be used, the installation program SWTCHSET has to be run from the DOS prompt before installing the interface card. SWTCHSET examines the configuration of our system and recommends the switch settings that should be used for the interface card. There is an option for us to print out the settings, if we need to.

SWTCHSET is loaded on to our hard disk automatically as part of the software installation process. To invoke the program, we first switch to the directory where the DeskScan software is installed. Normally this will be the default recommended in the installation process, DESKSCAN. We then type SWTCHSET, and follow the instructions that appear on the screen. Once we have the details of recommended switch settings, we can set the dip switches accordingly.

Step 3

Now the card is ready for installing, so the first thing we do is to turn off our computer. It is a good thing to remove the mains lead from the system unit as well—safety first! We can then open up the computer, removing the outer casing (saving the screws) and exposing the system board and accessory slots. The scanner interface card is an 8-bit card, so any empty slot will do. Remove the slot cover from the one chosen by removing the screw—and keep it safe. The card can then be inserted gently but firmly, and secured with the screw removed from the slot cover. Replace the system unit cover, though it is probably best not to secure it with all the screws until we know for sure that the card is correctly installed, and don't need to get back inside to readjust the dip switches. Finally reconnect the mains lead.

Step 4

Attach the small end of the interface cable to the interface card, and the larger, SCSI, connector to the scanner. Then, when we have also attached the scanner mains lead, we are just about ready to test the scanner. First, if we have gone through the steps above immediately after unpacking the scanner, we have a final task—to unlock the scanner. All flatbed scanners have a locking mechanism that secures the scanner's delicate innards—especially the mirror assembly—during transit. Any time the scanner is to be moved, the locking mechanism must be set to LOCK (Fig. 6.3). The user manual provides details on how to unlock a particular scanner.

Step 5

At this stage of the installation procedure we can test whether the interface card and the scanner have been installed correctly using the Scantest program,

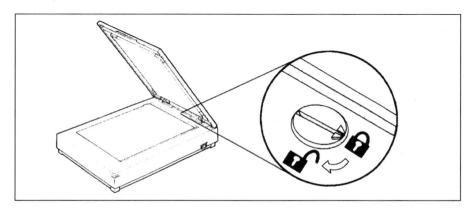

Figure 6.3 Transporting the ScanJet IIp

which can be started from DeskScan or from Microsoft Windows. Start by
powering on both the PC and the scanner, then start up Windows 3.1, select
the DeskScan II icon, and double click on it to start up the test.When we have
clicked on OK in the *Scanner Test* dialogue box, the test proceeds (Fig. 6.4). If
the test is successful, then we are told quite clearly (Fig. 6.5). If the test was
not successful, then we need to return to the user guide, where we will find
guides for troubleshooting. After the test has been completed successfully we
can start scanning using DeskScan.

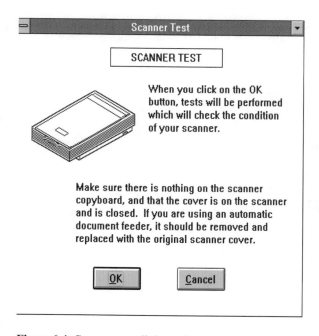

Figure 6.4 *Scanner test* dialogue box

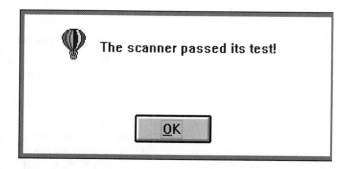

Figure 6.5 Successful scanner test

SCANNING WITH DESKSCAN

DeskScan is a Windows application and has the file name DESKSCAN.EXE.
Start the program in Windows 3.1 by double clicking on the DeskScanII icon.
Once DeskScan has started up, the screen shows a series of controls and a
scanning window (Fig. 6.6).

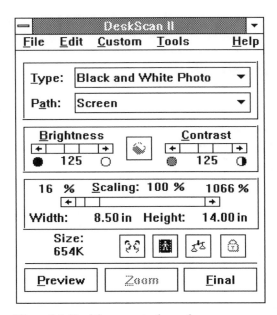

Figure 6.6 DeskScan control panel

SCANNING LINE ART

In order to produce a scan using this system, we first place a line art image,
face down, in the scanner (Fig. 6.7) and close the document cover, make sure
we have switched the scanner on, and load DeskScan.

Before we go ahead and scan, we need to think a little. What kind of
document are we about to scan? Is it line art or is it a photograph? These
questions are important. They help us to decide which settings to make on the
control panel. If we start getting into the habit of thinking in this way right
from the start, it will save us time later. Typically, the control panel will be set,
by default, to scan at the resolution of our screen—75 dpi—and if we go ahead
at that setting for line art subjects, then we will be very disappointed.

DeskScan can be a little confusing when we have been using other scanning
software. Where is there reference to scanning resolution? Well, it's not there
on the control panel explicitly. We set the resolution by means of *Path*, and
that's where the confusion creeps in. To many PC users, 'path' simply refers to

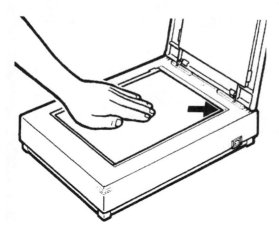

Figure 6.7 Placing an original in the scanner

the place a file is to be put, or from where it can be accessed. *Path*, in the DeskScan context, means 'target resolution', and as such is expressed in terms of *Screen, LaserJet, PageMaker/LaserJet* and so on, in other words, in terms of the output device. It is quite logical really, and it is a sign of how companies such as Hewlett-Packard are attempting to simplify the technicalities of scanning. Broadly speaking, the dpi set by *Path* will be 75 for *Screen*, 300 for laser and monochrome inkjet printers, and 180 for colour printers. We can change these values if we wish, or add in new 'paths'. For this test, we will select one of the supplied 300 dpi settings—*PageMaker/LaserJet*. Next we choose the kind of image we are to scan, using the *Type* box. This covers a range of image types, ranging from *Black and White Drawing* to *Black and White Photo*. We will select *Black and White Drawing*. Then click on the option *Preview* in the control panel of the activated *Scanner* window and the scanner will start up. Once a preview pass has been made across our image, a low resolution version will appear on the screen (Fig. 6.8).

So far, the scanner has simply made a fast pass across the image, and this is called a prescan. Frequently, we will not want to scan the whole of the A4 sized area which is available to us, and the view which is shown to us after the prescan enables us to select a particular part to scan. This makes the process of scanning much faster, and means that we will get a smaller scanned image file. We choose what we want to scan using the selection frame which is positioned by clicking and dragging the mouse pointer over the image in the *Scanner* window (Fig. 6.9). The scanning control panel on the screen allows us to vary the size of the image we are scanning: we can increase, or decrease its size. Changing the width of a scan will automatically cause the height to be adjusted, in proportion, as DeskScan keeps the original aspect ratio. For this scan we will use the default settings and when we click on *Final* another window is

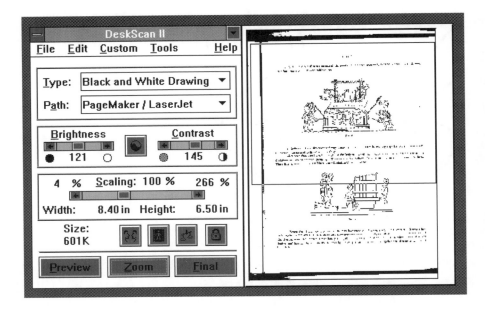

Figure 6.8 The scanned line art original in the *Scanner* window

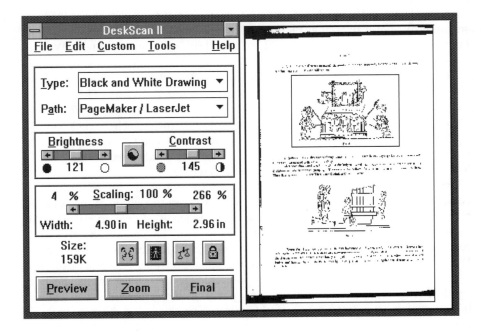

Figure 6.9 Selection frame in the *Scanner* window

opened to allow us to tell the program where to save the scan (Fig. 6.10). Once we have typed this in and pressed ENTER, or clicked on OK, the scanner will make another pass across our image, saving it in the file we specified. As well as defining where we want our scan to go, we can also tell the program the format in which we want the file saved. If we choose PC PaintBrush format, then the default name will be UNTITLED.PCX, but we can change this to whatever we want. PC PaintBrush format is a very useful way to save line art images, as we can import them directly into PC PaintBrush itself for editing.

DeskScan gives us a small image processing capability. It is mostly designed to scan a range of images, perhaps to re-size or adjust brightness and/or contrast, and to enable us to pick which portion of the A4 scanning platen we want to scan. We can also print our images directly from DeskScan, though it is more usual to use another application for this purpose. Hewlett-Packard supply PhotoFinish for more sophisticated manipulation.

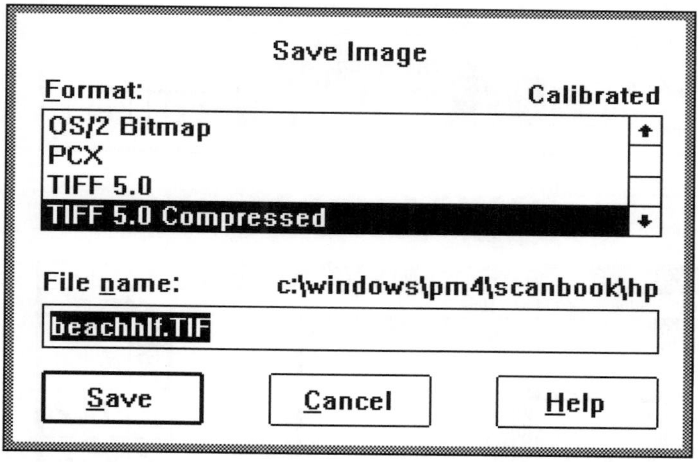

Figure 6.10 Saving the scan

EDITING LINE ART USING PHOTOFINISH

We move to PhotoFinish when we want to edit our scans. When we have double clicked on the PhotoFinish icon, the *PhotoFinish* window opens (Fig. 6.11).

In the normal way, we can *Open* our scanned files using the FILE menu. Once an image has been brought into PhotoFinish, we can zoom in and out of it using the magnifying glass option from the tool box (Fig. 6.12). With this tool, the left-hand button of a two-button mouse zooms IN, and the right-hand button zooms OUT.

Images that have been scanned can be cleaned up by zooming right down to pixel level, where a pixel-by-pixel edit may be made, or combinations of pixels may be changed simultaneously (Fig. 6.13).

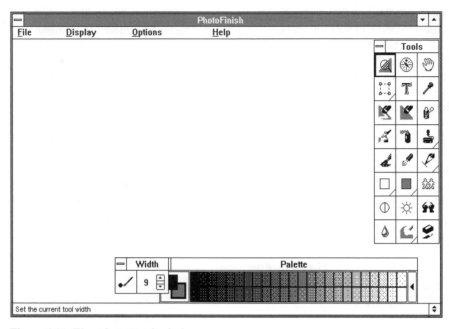

Figure 6.11 The *PhotoFinish* window

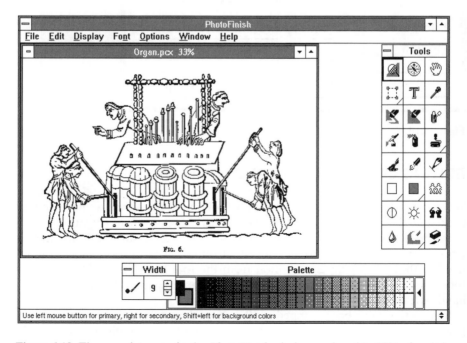

Figure 6.12 The complete scan in the *PhotoFinish* window, reduced to 33% of real size

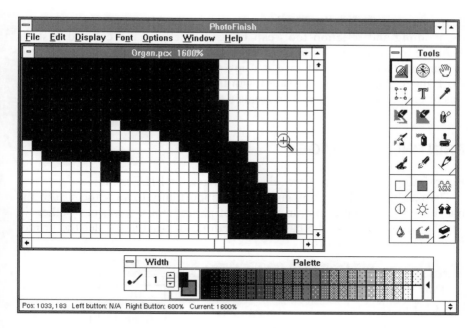

Figure 6.13 Zooming in to pixel level

A particular problem that happens when we scan line art images from less than perfect originals is that we have to spend time removing blemishes and spots from our image once it has been saved. Normally we might do this, for example, using PC PaintBrush, or PhotoFinish, and working at pixel level clearing each blemish manually. There is, however a very useful feature of PhotoFinish that will automatically remove spots of varying sizes from a scanned image (Fig. 6.14). PhotoFinish provides us with a substantial range of tools, and we will describe them in more detail when we come to scanning photographs, when they really come into their own. They can still be extremely useful for line art, though. We have already seen how we can use the *Zoom* tool to edit line art at the pixel level. The ability to move the scanned

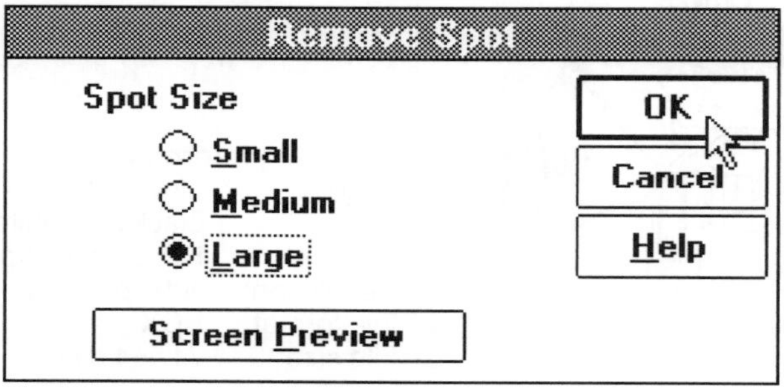

Figure 6.14 Spot removal

image on screen using the *Hand* tool saves us using the scroll bars. Many of the painting and selection tools can be used to effect, and cut and paste is available to move portions of pictures about. When the image has been edited to our satisfaction, we can save it using the option *Save* or *Save As* in the FILE menu. The scan may be saved with the file name we gave it at the time it was originally scanned, or we can create a second file.

PRINTING THE IMAGE

For printing a scan directly, we select *Print* from the FILE menu.

The resulting dialogue box gives us the opportunity to vary the size of the image we have scanned and, in the case of photographs, to select whether we let PhotoFinish choose the halftoning method, or whether we let the printer choose (Figs. 6.15 and 6.16)

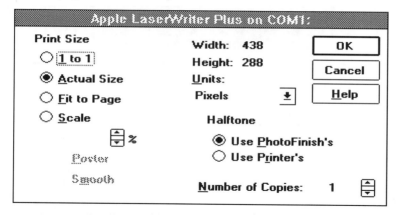

Figure 6.15 Printing from PhotoFinish

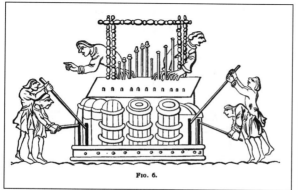

Figure 6.16 Printed scan of a line art original

SCANNING HALFTONES

The process of scanning a halftone original differs little from scanning line art. We place a halftone original such as a black and white photograph face down on the scanner and close the document cover. We start up DeskScan, but instead of selecting *Black and White Drawing* as the type of scan, we choose *Black and White Photo* or *Black and White Halftone*. We then select our *Path*, normally as the kind of printer we have, for example, *PageMaker/LaserJet*, which will have the scan resolution set to 100 dpi for grey scales of photographs, or, if the scan is being made as a halftone, at 300 dpi which is used for line art (Fig. 6.17).

Figure 6.17 The scanned grey scale original in the *Scanner* window, with default *Path* selected

CALIBRATING A PRINT PATH

If we find that our printer is not in the list defined for *Path*, then it is very much in our interests to put it there. In addition, if we are usually going to print our scans via a particular application, then we should ensure that is taken into account. Figure 6.18 demonstrates the difference in printed image quality between an inappropriately calibrated path and one that we have specifically set up for our needs. In the case of the left-hand image we scanned a photograph using the *PageMaker/LaserJet* option that comes as default. We were not very pleased with the result. We then calibrated the path for our Apple LaserWriter printer and rescanned the image, producing the right-hand picture which is very much better!

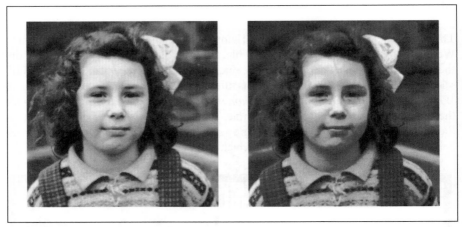

Figure 6.18 left, uncalibrated; right, calibrated for PageMaker and LaserWriter

We can calibrate a print path in two ways: first, by calibrating for the printer, second, by calibrating both for the printer and for the application or applications we are going to use. In either case, there is a very useful on-line help facility to show us what we need to do (Fig. 6.19).

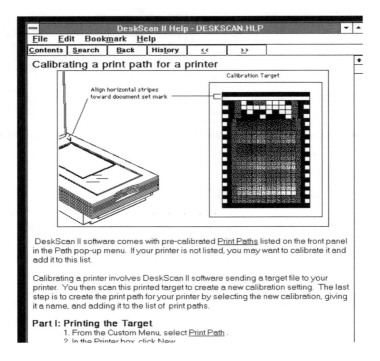

Figure 6.19 DeskScan help for calibrating a path

We can either print out the help topic, or leave it on the screen as a prompt.

The process of calibrating is quite straightforward, and consists of the following steps:

Calibrating a printer	**Calibrating an application**
1 Print a calibration target directly.	Print a calibration target to file. Close DeskScan. Open application to be calibrated. Import the target. Print it via the application.

The rest of the operation is the same whether calibrating a printer, or an application.

2 Place the target on the scanner. Click OK to scan.

When successfully completed, a *Save Calibration* dialogue box will appear. Click *Save*. Click *Quit*.

3 Add the print path to the list.

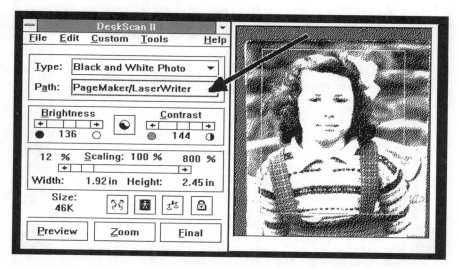

Figure 6.20 Grey scale scan in *Preview* window with newly calibrated path selected

Once the new path is there, we should always use it when scanning for the particular application. So which setting do we use for our photograph? Grey scale—or halftone? Broadly speaking, we found grey scale to be the most effective, and grey scale should in any case be used:

1 When any image manipulation is needed over and above fairly basic rotation, cropping and so on. Such manipulation could be scaling, blending, adjusting for contrast and so on.

2 When we are going to print on a phototypesetter. There are other advantages to scanning in grey scale mode. For a start, the image on screen will be closer to the original.

Carrying on with the scanning process, we now delineate the part of the image we want for the final scan, in exactly the same way as we did for the line art scan. When the required area is defined, we choose the option *Final* from the SCAN menu. The dialogue box appears so that we can choose how we want the scan saved (TIFF, PCX, etc.) and how we want to name it. If we choose to scan as a halftone, then there is a range of halftone types that may be selected via the CUSTOM menu (Fig. 6.21). The range covers *Diffusion, Normal, Fine, Extra-Fine, Horizontal Line* and *Vertical Line*. Each of these is beneficial with a given range of circumstances. For example, a *Diffusion* halftone is best used where a previously halftoned image is to be scanned— maybe a picture from a magazine. It minimizes the interference patterns that may otherwise ruin the scan. A *Normal* halftone photocopies well.

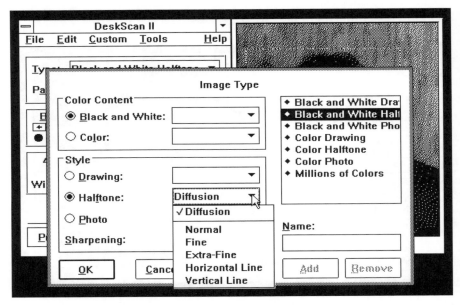

Figure 6.21 Selecting a halftone type

SCREENING HALFTONE ORIGINALS

Using halftone screens, we can give the impression of shades of grey when we display an image on a monitor, or send one to a printer. Such screened images can reproduce black and white photographs quite acceptably, and screens should always be selected when an application program for the further use of halftones scans (i.e. Word for Windows or WordPerfect) does not support the import of scans with real grey shades. When using a desktop publishing program such as PageMaker we should avoid producing screened scans especially if they may need to be resized. In DeskScan there are six different screens available (see Figs 6.24 to 6.29). These may be compared to the preceding Figs 6.22 and 6.23, which show the same image scanner and printer as 256 grey scales. Fig 6.22 was scanned at original size (100 per cent), then enlarged. Fig. 6.23 was enlarged in scanning to 400 per cent, then printed.

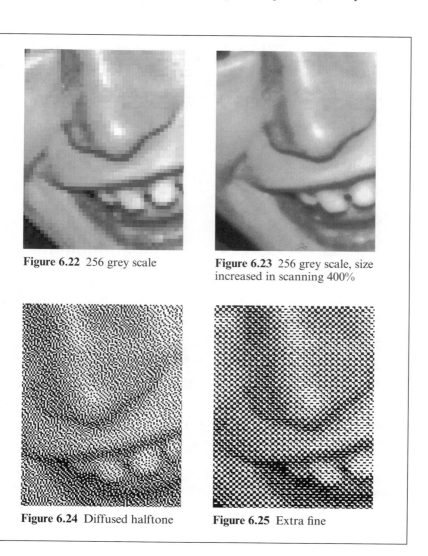

Figure 6.22 256 grey scale

Figure 6.23 256 grey scale, size increased in scanning 400%

Figure 6.24 Diffused halftone

Figure 6.25 Extra fine

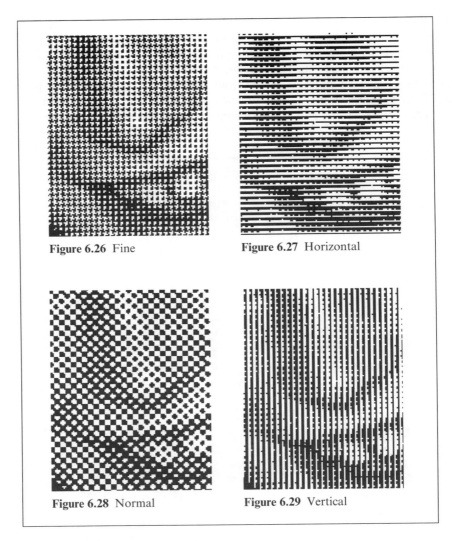

Figure 6.26 Fine   **Figure 6.27** Horizontal

Figure 6.28 Normal   **Figure 6.29** Vertical

Scanning Information
Original: Halftone, black and white
Scanner: HP ScanJet IIp
Resolution: 300 dpi
File format: TIFF compressed
File size: 18391 to 55608 bytes

FILE FORMATS

We can save scanned images in formats other than the default of TIFF. We are invited to select our scan format when we make a final scan of our image, and the dialogue box shown in Fig. 6.30 appears. The list of formats that may be used for image files are those for Microsoft Windows 3.1 Bitmaps (BMP), PC PaintBrush (PCX), GEM, EPS (Encapsulated PostScript), EPS for screen image, OS/2 Bitmap, in addition to TIFF 5.0 and TIFF 5.0 compressed.

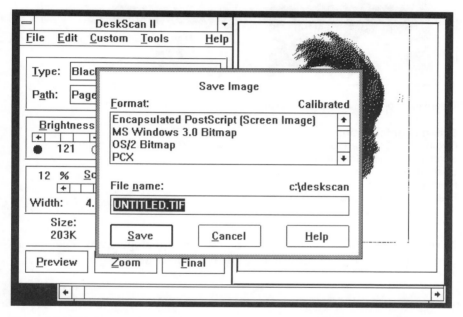

Figure 6.30 File formats in DeskScan

FURTHER OPTIONS

In the line art, grey shade or screening modes in DeskScan there are various options available and we discussed their impact on a scan in Chapter 3 'Resolution, grey steps and screens' and in Chapter 4 'Originals for Scanning'. Let us look at the options briefly.

Negative image When we select the option *Negative Image* icon ( ), we will get a negative or colour inverted version of the original. We can also select this option at prescan time. Figure 6.31 shows an example of a negative scan.

Figure 6.31 positive and negative scans

Scanning Information
Original: Line art, black and white, reduced to 50 per cent of original size
Scanner: HP ScanJet IIp
Resolution: 300 dpi
File format: TIFF
File size: positive: 34 167 bytes
                negative: 33 736 bytes

Mirror Image This option in the *Image Scan* dialogue box enables a mirrored scan (🔲) of the original. Fig. 6.32 shows an example.

Scanning Information
Original: Line art, black and white
Scanner: HP ScanJet IIp
Resolution: 300 dpi
File format: TIFF
File size: 34 1247 bytes

Figure 6.32 Mirrored scan

Grey scale When we select this option as a *Type*, the original will be scanned unscreened, that is, with different grey values or with different grey shades. If the image is to be used later in other software (with PageMaker, for instance) we must choose this option, which has a choice of 16 grey shades (4-bit grey scale) and 256 grey shades (8-bit grey scale).

Grey scale

The expressions grey steps, grey shades and grey scales can be used interchangeably. Figures 6.33 to 6.38 show results using a 300 dpi laser printer to print a halftone original, which was scanned at different resolutions, scalings and grey steps.

A comparison of the images in Figs 6.33 to 6.39 shows relatively small differences for the different grey steps, resolutions and scaling factors. The main reason for this is the 300 dpi laser printer which shows its limitations while printing screened halftones at 74 dpi and with 64 grey steps. The differences are more marked when printing with a type- or imagesetter on photographic paper.

Figure 6.33 256 grey steps, 100 dpi
TIFF file size: 18346 bytes

Figure 6.34 16 grey steps, 100 dpi
TIFF file size: 8124 bytes

Figure 6.35 256 grey steps, 100 dpi, enlarged 150% in PageMaker
TIFF file size: 18 346 bytes

Figure 6.36 16 grey steps, 100 dpi, enlarged 150% in PageMaker
TIFF file size: 8 124 bytes

Figure 6.37 256 grey steps, 100 dpi, enlarged 150% in the scan program
TIFF file size: 38 110 bytes

Figure 6.38 16 grey steps, 100 dpi, enlarged 150% in the scan program
TIFF file size: 15 974 bytes

Figure 6.39 Importing a scanned image file to PhotoFinish

MANIPULATING GREY SCALE IMAGES

Figure 6.39 shows a photograph that has been scanned using DeskScan, saved as a TIFF file and opened in PhotoFinish. This screen dump highlights very well the effectiveness of Windows 3.1 as a working environment. Frequently when we are scanning for a particular purpose (as opposed to experimenting) then we can usefully need to have a range of applications open at the same time. Here, we have PhotoFinish as the active window, DeskScan remains open for additional scanning as needed, PC PaintBrush is also open to edit images when needed, as is Aldus PageMaker, the application to which the final scans are being imported. DeskScan and PhotoFinish each permit us to sharpen an image. In the case of the image in Fig. 6.39, we elected to do the sharpening in DeskScan (Fig. 6.40).

If we had not made the sharpening adjustment during scanning (Fig. 6.41), then we could still do it once the image is in PhotoFinish (Fig. 6.42).

Figure 6.40 Sharpened normal

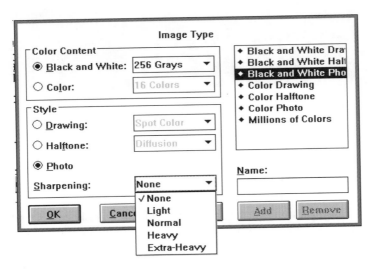

Figure 6.41 Sharpening an image during scanning

So the image is now on the screen. How can we change it? What can we do with it? The answer is pretty nearly anything we want to, given the tools at our disposal. Not everything will give us sensible results, however! Many of the things we are able to do can result in horrendous end products, so it is best to start in a limited and simple way, until we have had the chance to explore just what will work successfully, and what will not. The picture in Fig. 6.43 shows how we can drop a box containing a graded fill and some text into a scanned

Figure 6.42 Sharpening the image after scanning, in PhotoFinish

Figure 6.43 Box and text added

image. This is very simple editing and we could do similar operations once we have imported the scanned image into PageMaker, but it is very easy to do it within PhotoFinish, and the change will then always be reflected in the saved graphic file. In PageMaker, we would have to drop a single colour box on to the image, and this could not be locked in place. Neither could the text we would have to drop in place afterwards. Sometimes we want to draw attention to something within a particular picture. We can do this by keeping that part distinct, while making the rest of the picture less distinct. In Fig. 6.44, we have used the blurring tool to blur out most of the picture except for the cat. Besides changing the light values, the sharpness and so on of the picture we scanned, we can also materially change its composition, by cutting and pasting. This we can do within a single picture, or from one picture to another. In Fig. 6.45 we have duplicated part of the original cat and superimposed the duplicated image on the original picture. We could have enlarged or reduced

Figure 6.44 Photograph selectively blurred within PhotoFinish.

Figure 6.45 Image cut and pasted

Figure 6.46 Using edge detect

the size of the duplicated image before placing it. Some of the image manipulation tools enable us to produce quite striking, albeit not realistic, effects. In Fig. 6.46, for example, we have used the EDGE DETECT capability. Only a short while ago, it would have been necessary to spend a substantial sum to buy a program with the capabilities of PhotoFinish. It is a measure of the increasing maturity of the scanning market that this program is now supplied as a normal part of the ScanJet scanning package.

IMAGE EDITING

Among the features that PhotoFinish gives us are:

Display tools:

- Zoom (changes magnification)
- Locator (for seeing selected area in multiple open pictures)
- Hand (moves picture around).

Selection tools:

- Box (defines rectangles)
- Magic wand (defines areas with similar colours)
- Lasso (defines irregular areas)
- Scissors (defines polygonal areas)

Painting tools:

- Includes tools to add text, select colours, erase, replace, spray, splatter, fill, draw shapes, lines, curves, polygons, clone areas

Retouch Tools:

- Ability to adjust contrast, intensity of colours, change shades, smooth, smudge, and sharpen.

Such a range of tools is daunting at first. However, with patience and experimentation, and the ability to resist the use of too many of the effects in a single image, many worthwhile results may be obtained. Some of the effects can be quite surprising, leading from photographic images to what might be regarded as abstract art. In Fig. 6.47, the left-hand image of bluebells we scanned earlier has been transformed using *Edge Detect* to produce the central image, which then had its contrast greatly enhanced. Other effects permit further manipulation of the image (see Figs. 6.48–6.50).

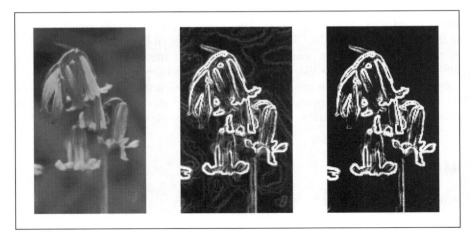

Figure 6.47 Changing an image from photo-realistic to abstract

Printing with Aldus PageMaker and Ventura Publisher

Once we import our scanned image into an application such as Aldus PageMaker we limit ourselves to adjustments for brightness and contrast. We can make such changes when we select our imported image, then choose the ELEMENT menu, and click on IMAGE CONTROL. This will give us the dialogue box in Fig. 6-51.

Figure 6.48 Image embossed   **Figure 6.49** Image
equalized

Figure 6.50 Image edge
detected

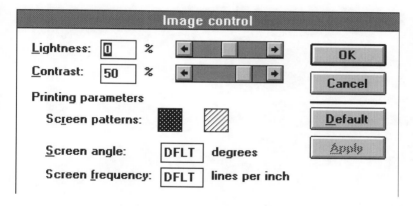

Figure 6.51 The PageMaker dialogue box for the enhancement of images

We are unlikely to need to change the setting for *Screen pattern*, unless we
want to create some special effect, and the *Screen angle* and *Screen frequency*
are automatically set by PageMaker to be the settings best suited for our
printer.

Scanning halftones from publications

When we considered halftone images, we said that *Diffused* halftone is the
selection to make when we scan an image that has already been halftoned,
such as a picture from a newspaper or magazine. Figures 6.52 and 6.53 show
the comparison between grey scale and diffused halftone representations of a
newspaper picture.

Figure 6.52 Grey scale representation of a halftoned image

Figure 6.53 Diffused halftone of a previously halftoned image

SUMMARY

In this chapter we have used a typical flatbed scanner and the software that is supplied as a part of the scanning package. The Hewlett-Packard ScanJet IIp is a powerful scanner as there are up to 256 grey shades available with a resolution of 300 dpi. The A4 scanning area and the software controlled scaling between 12 and 800 per cent are important additional features. The only drawback—and this is common to all desktop scanners—is the potential for damaging bound books when scanning from them. It is all too easy to break their spines when laying them on the scanning platen. This same problem affects photocopiers as well, of course. DeskScan, which is supplied as a part of the scanner package, has a number of useful capabilities for simple image processing while scanning. The inclusion of the very powerful PhotoFinish software makes the ScanJet IIp package a very powerful one indeed. Of particular note with the ScanJet package is the largely successful attempt to demystify the business of getting successful scanned images. The calibration of print paths is particularly effective, and the ease with which images may be manipulated is quite stunning.

Chapter 7
Resolution, File Size and File Formats

The images we scan will very quickly eat up the space on our fixed disk, and one of the first lessons we learn is that we need to have a large amount of space where we can store our scans. The scan programs offer us a variety of options to enable us to optimize the use of our file space: for example, we can choose the resolution, number of grey steps and file format for our scans. All of these affect the size of our saved scan; some of them also have implications for the use of the RAM on our computer system.

IMAGE SIZE

As we look at the original we want to scan, we need to appreciate that the amount of scanned information increases in direct proportion to the size of the original image. This is as valid for the space the scan takes up in the main memory of our computer (depending on the type of scan program) as for the file we will save to store the scan. The graph in Fig 7.3 shows this relationship for the line art and halftone examples in Figs 7.1 and 7.2. To clarify this, we have scanned three different-sized originals. We scanned the line art at 300 dpi, the halftone at 75 dpi with 16 grey steps at 1:1 scale, and saved the scans in TIFF format.

Figure 7.1 Line Art (from CorelDRAW)

Figure 7.2 Halftone original

We then scanned identically sized line art and halftone originals using the HP ScanJet IIp flatbed scanner at 100, .50 and 25 per cent of original size. The line art was scanned at 300 dpi each time and the halftone original was scanned at 75 dpi with 16 grey steps. The graph in Fig. 7.4 shows the relationship between scan size and file size. When we compare both graphs we can immediately see that the different sizes of the originals in Fig. 7.3 correspond to the scaling percentages in Fig. 7.4.

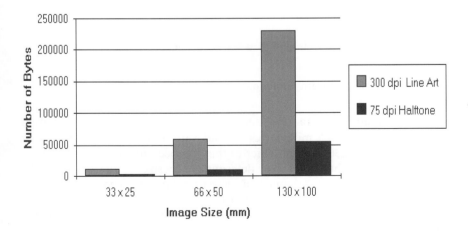

Figure 7.3 Image size and bytes used

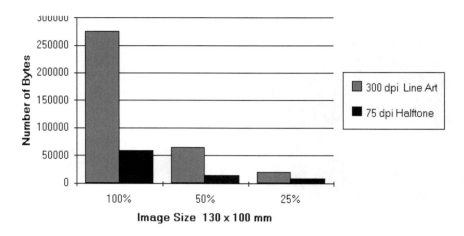

Figure 7.4 Scaled scan and bytes used

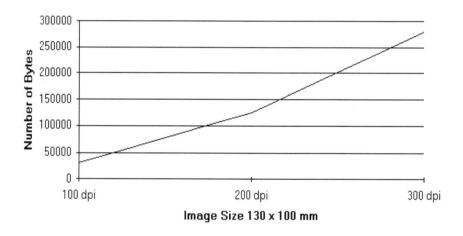

Figure 7.5 Relationship between resolution and file size

When we compare the data presented in Figs 7.3 and 7.4, we can see that an original image of 33 × 25 millimetres will take up a very similar amount of space to an image of 130 ×100 millimetres that has been reduced to 25 per cent of its original size. The image size determines the size of the scanned file, regardless of whether the original was a particular size when scanned, or whether it was enlarged or reduced in size during the scanning process to a particular size. The file size of a line art image scanned at 300 dpi is roughly four times larger compared with a halftone scanned with 75 dpi and 16 grey steps, as shown in Figs 7.3 and 7.4.

RESOLUTION

In Chapter 3 we learned about the relationship between resolution, screen and print quality. Now we will examine relationships of the resolution and the file size of a scan. To make this clearer, we scanned the line art from Fig. 7.1 at a size of 130 × 100 millimetres at resolutions of 100, 200 and 300 dpi, and the scans were saved as TIFF files. The chart in Fig. 7.5 shows that the file size increases in proportion to the resolution chosen: each doubling of the resolution increases the file size by a factor of four. Where an application cannot itself apply a screen to a scanned halftone original, the image must be screened by the scan program before sending it to a printer of 300 dpi or less. If we are using a printer such as the HP LaserJet, we must scan halftones at a resolution of 75, 150 or 300 dpi. After scanning, we save the image in a screened format. The halftone original from Fig. 7.2 was scanned in this way, and each scan was saved in a TIFF format file. The graph in Fig. 7.6 shows the relationship between resolution and file size of these screened halftones. As with line art, each doubling of the resolution increases the file size by a factor of four. In general, we can assume that the higher the resolution of a scan, the higher the quality when it is printed, as long as the scan resolution does not exceed the printing resolution of the printer being used. So far we have seen that the image size is of major importance in determining the size of the original when it has been scanned and saved. Image size is not the only determinant of size, however, the following discussion of TIFF format images shows that the number of grey steps also has an effect on file size.

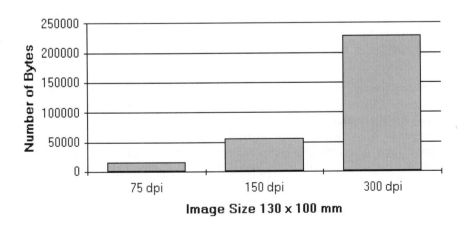

Figure 7.6 Relationship between resolution and file size

TIFF FORMAT

We saw in Chapter 3 ('Resolution, grey shades and screens') that image files with grey shades are normally printed with halftone screens of 53 lines per inch for printers with a resolution of 300 dpi. When printing with this screen width 75 dpi will be the best resolution for the scanned halftone original. By doubling the resolution in some cases we may get a better image quality, but, as we have seen from the tests described above, this leads to a file size that is four times larger. The file will take very much longer to print. We save half-tone originals with grey shades as TIFF files when we are using a printer with a resolution of more than 300 dpi, or a typesetter such as a Linotronic, and we calculate the scanning resolution by multiplying the halftone screen width (in lines per inch) by the square root of 2, Table 7.1—this is taken from the Hewlett-Packard ScanJet Plus (an earlier version of the ScanJet IIp) user manual—may help us to make our choice. The listed values are based on an image size of 200×250 millimetres.

Table 7.1 Screen width, resolution and file size

Halftone screen width (lines per inch)	Minimum scan resolution (dpi)	File size 256 greys (8-bit)	File size 16 greys (4-bit)
200	283	6.4 Mb	3.2 Mb
150	212	3.6 Mb	1.8 Mb
133	188	2.8 Mb	1.4 Mb
120	170	2.3 Mb	1.2 Mb
100	142	1.6 Mb	800 Kb
90	127	1.3 Mb	650 Kb
80	113	1.0 Mb	510 Kb
60	85	580 Kb	290 Kb

Output devices with resolutions that are far higher than 300 dpi are used by professionals. Pictures in newspapers, for example, are printed with screens of 60 to 100 lines per inch. Magazines normally use finer screens of around 150 lines per inch. Table 7.1 also shows that a halftone whose size is 200×250 millimetres, scanned with 256 grey shades and with a minimum resolution of 212 dpi, will require 3.6 Mb when saved in TIFF format. The same original scanned with 16 grey shades would still result in a TIFF file of nearly 1.8 Mb. These file sizes clearly exceed the storage ability of even the highest capacity floppy diskette, 1.44 Mb on 3.5 inch media. If we want to send such large files to a professional typesetter, then we will need to have some appreciation of the means available for archiving, compressing and distributing such large files.

Some scan programs, such as Hewlett-Packard's DeskScan can store scans in a compressed TIFF format. A compressed file takes less space on a disk but needs a longer time for loading. Before saving a scan in a compressed format for use in another application, we need to be sure that the application can read such files. The desktop publishing program Aldus PageMaker will, for example, work with compressed TIFF files without any problems. Figure 7.7 shows some line art scanned 1:1 at a size of 110 × 75 millimetres, and at 300 dpi. This scan was saved with the normal TIFF format (153 729 bytes) and also with the compressed TIFF format (57 566 bytes): the compressed file took only 30 per cent of the space of the normal TIFF format.

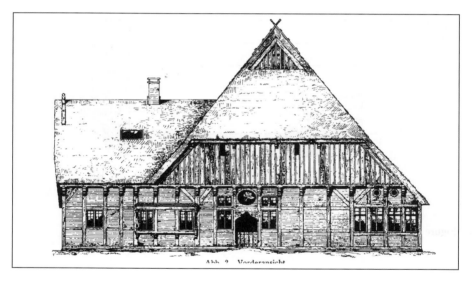

Figure 7.7 Line art saved in normal and compressed TIFF format

Scanning Information
Original: Line art, black and white
Scanner: HP ScanJet IIp
Resolution: 300 dpi
File format: TIFF
File size: 153 729 and 57 566 bytes

EPS FORMAT

Some scan programs permit us to save scans in EPS (Encapsulated PostScript) format. EPS format has become a standard in desktop publishing applications that use a PostScript printer or typesetter, but we should use it only rarely for storing scanned originals. The main reason for this is the extremely large file size as shown in Fig. 7.8. The figure shows the relationship between file format and file size. A 130 × 100 millimetres sized halftone original was scanned 1:1 at 75 dpi and stored in TIFF, PCX and EPS formats.

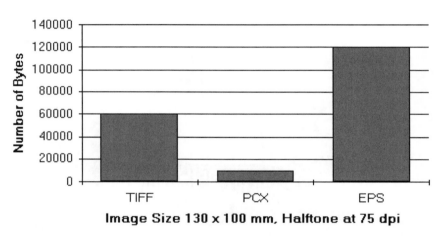

Figure 7.8 A comparison of saved file sizes for the same image

PCX FORMAT

Saving in PCX format can be an advantage if we want to edit the scan at a later stage using paint programs such as PC PaintBrush in both DOS and Windows 3.1 versions. The PCX format is very widely used as a means for graphics interchange, and for storing graphics images generally. PCX format files may be imported into many applications, including word processing, desktop publishing and graphics, and scanned images may be printed in this way from these programs.

SUMMARY

When we scan we create large image files. The exact file size will depend on the original image size, and the enlargement or reduction that is applied, the resolution chosen and the number of grey steps. We can save our scan in different file formats. The format we choose depends on the application that we will use for editing, printing or for combining scan and copy for the scan.

Chapter 8
Scanning with Colour

Black and white scanners, especially those that handle halftones, provide just about all of the facilities and quality that we will need for most practical, everyday work. To get the most out of scanners that can use grey scales and resolutions in excess of the 300 dpi most commonly used for laser printers, access to a typesetter such as a Linotronic is essential. As early scanners moved from being able to capture simple line art, to being capable of handling shades of grey, all at a wide variety of resolutions, so manufacturers are now selling colour scanners. There is an increasing number of colour scanners available at the moment, and they not only scan in colour, but in black and white halftone mode (300 dpi and more with up to 256 grey steps) as well. Scanning itself is, of course, not the entire story: we usually want to print the objects we have scanned. For black and white images this is relatively simple as most PC systems have some form of monochrome printer available to them, and these, generally, will make a very good reproduction of the image that has been scanned. With a colour scanner the solution is not as simple, as few PCs yet have a colour output device available to them. Those that do are likely to have colour dot matrix or inkjet equipment. The latest colour inkjet printers, such as the Hewlett-Packard DeskJet 500c can provide a very convincing rendition of a scanned colour original. Colour laser printers have begun to appear on the market, but they are variable in output quality (they tend to use specialized techniques of printing), and are very expensive. At the moment some of the best results are obtainable from thermal transfer printers, which print the three basic colours—yellow, cyan and magenta—in a screen pattern with three runs on special paper or film. The final result is a coloured print of the scanned original. Colour image processing using a PC is in its infancy, and few installed PC systems can handle image sizes of up to 20Mb for a single colour A4-sized original. Even using a power system with a 80486 processor, cache, high speed fixed disk and fast VGA graphics cannot overcome the relatively slow processing of a colour image. Printing colour at high quality can also be three times slower than printing an equivalent monochrome image, because of the need to make a print in several passes.

In spite of the problems, colour image processing is generating more and more interest. At present there is no sign of a software and printer standard for colour processing, but progress will accelerate with time. It is probable that the developing area of multimedia will bring everything together. Scans produced using Microtek MSF 300—Z and EyeStar II scan software may be saved in a colour PCX file that can be imported into Designer 3.0 for editing in colour line art mode. There are already page layout and graphic programs available that can process colour data and send output to a printer on paper or film. Colour scanners are in common use at typesetting studios for creating colour separations.

MICROTEK COLOR/GRAY SCANNER MSF—300 Z

The Microtek flatbed image scanner MSF—300 Z (see Fig. 8.1) can scan a colour image and produce a 24-bit code indicating any one of 16 million colours. The scanner can also work in black and white scanning mode with an 8-bit code for up to 256 levels of grey. With this scanner a user has access to all established black and white applications (image and graphic processing in line art, halftone or grey scale mode, character recognition). For someone who wants to buy a scanner now for monochrome scanning, without cutting out options for colour processing in the future, then this scanner represents the current state of development of colour scanners. The MSF—300 Z is supplied with EyeStar II scanning software (see later in this chapter).

Figure 8.1 Microtek MSF—300 Z

The Microtek scanner works with one bulb and scans the original in three runs by changing between three colour filters. Other colour scanners, such as the Sharp, use three coloured bulbs in a single pass. Working with a single bulb has advantages in conventional scanning with grey shades, as the scanner uses white light to avoid generating false information. The one bulb has a much longer lifetime because the coloured bulbs used in some other scanning systems are repeatedly switched on and off when scanning coloured originals.

The maximum resolution of the MSF—300 Z is 300 dpi with a document size of 8½ × 14 inches (216 × 356 millimetres). In what is called multiple bit or 8-bit mode it is possible to scan an original with up to 256 values of grey. The software scales a scan between 25 and 400 per cent on the x- and y-axes. These figures are comparable with those of other black and white or grey shade scanners such as the Hewlett-Packard ScanJet IIp or the Microtek MSF—300 G.

As this book is printed in black and white we cannot show what a colour scan looks like. Our work with colour scanners has proved that using quality colour scanning software the scan on the screen comes very close to the original. In colour mode the MSF—300 Z can recognize up to 16.8 million colours in three single runs (red, yellow, blue). EyeStar software enables us to control lightness and contrast for these three basic colours. Scaling is set up between 25 and 400 per cent and the resolution between 75 and 300 dpi. Because the original is scanned in the three basic colours in colour mode, the data depth is 3 × 8 bits per pixel or 24 bits per pixel. This is why the file size is three times larger than an equivalent image scanned with 256 grey steps. The scanning process for an A4-sized colour original takes, in all, 150 seconds with the MSF—300 Z in 24-bit mode.

EYESTAR II

The MSF—300 Z is supplied with EyeStar II, a scanning program that runs under Microsoft Windows. Besides controlling the scanner the software offers many features for editing the coloured images. Figure 8.2 shows the window for controlling the scanner. Before starting the scanning process we must use this window to set up the attributes or options for resolution, scaling factor, image type (colour or black and white), data depth and scanning velocity. There is also an option to change the weight of the three basic colours, enabling us to scan with more or less red, yellow or blue. All settings can be saved and used for later scans.

For the final scan we roughly predefine the scanning area, make a prescan and define the image area we want. After choosing the option *Scan* we will see the screen shown in Fig. 8.3. The upper right window tells us about the scanning status. The message '33 % Processed' tells us that one third of the image, or the first of the three basic colours has been scanned. Clicking on the *Abort* button stops the scanning process. Once the scanning process is

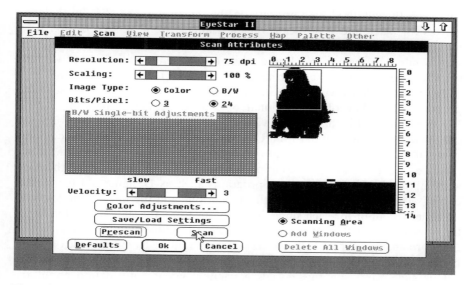

Figure 8.2 EyeStar II window for scanner control

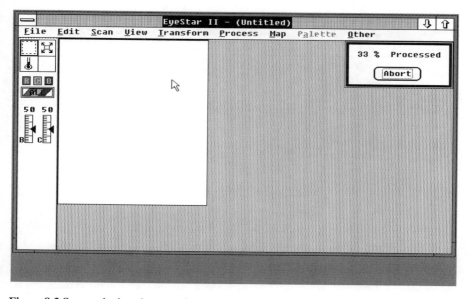

Figure 8.3 Screen during the scanning process

complete, the coloured scan appears on the screen (see Fig. 8.4), in the same way as for black and white scans on the scanners we have seen before. If we had chosen the option *B/W* (black and white) in the scanning window, the scan would be shown in black and white on the screen (this is what we have shown, since there are no colour images in this book).

The menu options within EyeStar II are similar to those in other scanning programs, but they offer additional options for colour processing. In the VIEW menu (see Fig. 8.5) we can view a scan in reduced or enlarged form. We can switch from the standard halftone mode (which we also use for coloured scans) to *Line Art View*; this shows a halftone scan in line art mode with the three basic colours and the resulting mixed colours. We could save this view and import the file into graphics programs such as Designer 3.0 for further editing. The colour balance in both the halftone and the line art modes is set up either for all or for each single basic colour using the slide bar to the left of the screen (see Fig. 8.5). We can change colour brightness and contrast.

Before changing the basic colours we may want to have some idea of their balance in different parts of the image. For this we select the object above the colour slide bars, and this reveals a window giving information about the colour values (0 to 255). As we move the cursor around the image, the colour values are shown (see Fig. 8.6).

After marking the image area that we want, using one of the different selection frames, we can cut, copy, insert, duplicate, invert, lighten and darken that part of the image using the options in the EDIT menu. When we invert a colour image EyeStar II shows a real colour negative on the screen. There is also an option to view the contents of the clipboard. Using the TRANS-FORM menu we can change the size of a scan or slant, rotate or mirror the

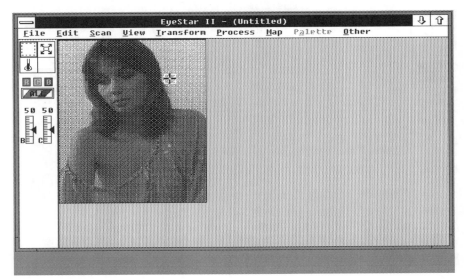

Figure 8.4 Screen with the coloured scan

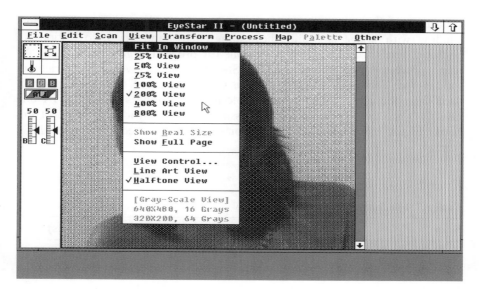

Figure 8.5 View control in EyeStar II

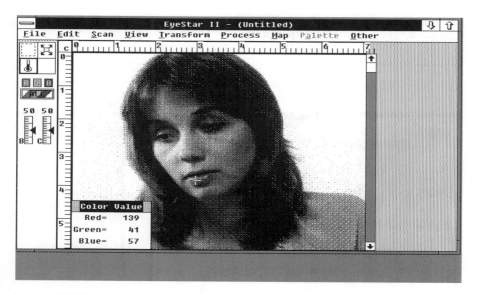

Figure 8.6 Displaying the colour values in EyeStar II

image. The PROCESS menu provides many of the filter functions that we will see in Chapter 9 with other programs that are available for image processing. Results for coloured scans are impressive, and not only for graphic designers. The MAP menu offers further options for image processing, enabling us to invert, optimize and posterize the scan as well as change lightness and contrast. Finally we can use rulers (inches or centimetres), an additional window with image size and the scaling factor and an undo option. In addition, EyeStar II offers many options for processing grey scale scans. All scans processed and edited with EyeStar II can be saved in a variety of formats (see Table 8.1). Formats such as GEM and EYESTAR are less widely used than the other 1-bit options. Colour, 8-bit formats like TARGA and GIOTTO are unique to colour processing and are outside the scope of this book. Figure 8.7 shows the window for saving a coloured scan in the TIFF format.

Table 8.1 File formats in EyeStar II

Scan	Export format	Data depth (bits)
B and W, 1-bit	TIFF, PCX, EPS	1
	GEM, EYESTAR	1
Grey, 8-bit	TIFF	1, 4, 6, 8
	PCX	1, 4
Colour, 3-bit	TIFF, PCX	3
Colour, 8-bit	TIFF, TARGA, GIOTTO	24
	PCX	3, 8

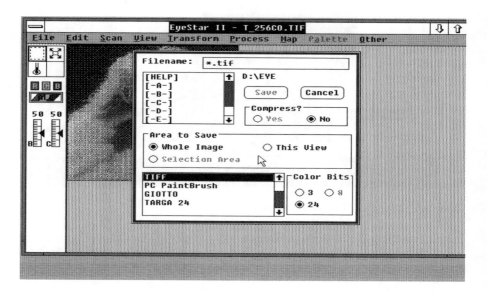

Figure 8.7 Saving a colour scan

OTHER COLOUR SCANNERS

We have already looked at the Logitech and Hewlett-Packard scanners, both companies have produced colour versions of their scanners and each one uses the same scanning and image manipulation software that is supplied with their grey scale scanners. The ease of use of their respective software applications leads to an extremely simple way to capture coloured images and manipulate them. Both of these scanners are also very competitively priced, and almost certain to gain the same hold over the colour market that they have over that for monochrome images.

SUMMARY

EyeStar II software, together with the Microtek MSF—300 Z colour scanner, is an excellent scanning and image processing package to scan, edit and process just about any black and white or coloured originals. For processing coloured scans we need a lot of patience, as we are dealing with three times more information than with black and white scans. To work successfully with EyeStar II we should have, at the very least, a fast 386 processor with a fast fixed disk, and a lot of memory. The FotoTouch and PhotoFinish software that are supplied with the Logitech and Hewlett-Packard scanners provide the same capabilities for image enhancement in colour as they do for grey scale images (see Chapters 5 and 6). Both scanners are also well recommended.

Chapter 9
Programs for Image Processing

In this chapter we will look at some of the programs that can help us to manipulate the graphics we have scanned. The first two, ImageIn and Gray F/X, can themselves drive a variety of scanners, besides being able to manipulate or process a scan. ImageIn has additional modules for vectorizing graphics and character recognition. Another program, Picture Publisher, can also accomplish advanced image processing. Lastly, CorelDRAW is a program that is well known for its drawing and type capabilities in addition to its ability to import externally produced scans. Scanned bitmaps may be combined with CorelDRAW line art graphics.

IMAGEIN

ImageIn is an integrated software package for scanning and processing documents, a Windows application that can be used for scanning line art and halftone originals, processing these images or scans, vectorizing scans, and for character recognition. ImageIn is supplied with Panorama, an image database that helps organize scans on a fixed disk. The flexibility of ImageIn, together with the wide variety of scanners it can drive, makes it a tempting choice in place of many of the programs that are offered with scanners that can only scan.

ImageIn and Windows

It is an easy task to install ImageIn within the Microsoft Windows environment, and it may be easily called up later as a Windows application. There are no problems in installing and setting up the software, as ImageIn works with any screen, printer and mouse that is supported by Microsoft Windows and selected by the user. The Windows environment gives ImageIn (and thus the scanner) the ability to interchange images, text and graphics with other Windows applications. An increasing number of Windows applications are available, including desktop publishing programs such as PagerMaker and Ventura Publisher; image processing programs like Picture Publisher and

Image Studio; drawing programs like CorelDRAW and Designer; word processing programs like Ami Professional and Word for Windows; spreadsheet/charting programs like Excel and Charisma; and database programs like SuperBase. The availability of standard file formats like TIFF, PCX and EPS enables us to export our documents and graphics to non-Windows applications. Windows allows us to have two applications, such as Windows Write and ImageIn, active at the same time. We can edit text, digitize an image with ImageIn, use this image within the text, scan a text document in ImageIn, convert this scan into characters and merge them with other text, all without having to close down any of the programs. The entire process may be done in a single working session without leaving Windows or one of the applications.

Features of ImageIn

In essence, ImageIn is a collection of powerful individual programs:

ImageIn The main program that controls the other modules. They all run under Microsoft Windows. The program loads and saves scans with different file formats, offers different image views and image information and provides the interface for the printer that has been selected with Windows.

ImageScan Module for digitizing or scanning line art and halftone originals. This is the module that can control a wide range of scanners, including those from Logitech, Canon, Hewlett-Packard, Panasonic, Microtek, Ricoh and others.

ImagePlus A program with all of the features of ImagePaint but with additional options to process images with grey values. There are filters for selected image areas, controls for contrast and brightness and many ways to manipulate pictures with controls for softness, blur, solarization, and editing of edges.

ImagePaint A drawing and paint program for editing scans — in line and grey scale mode.

ImageRead A module for optical character recognition. Scanned text may be stored in ASCII format or transferred into the Windows clipboard. During the recognition phase ImageRead can learn to interpret the characters it is reading.

ImageVect A module for converting bitmapped graphics into vectorized images. These may then be edited further in drawing programs like CorelDRAW or Designer.

Panorama Panorama is an independent Windows application for managing picture libraries via keywords. The program can be used for picture files in TIFF, MSP, PCX, and MacPaint formats.

Though it is possible to obtain the ImageIn modules separately, some modules are dependent on others: for example, ImageRead depends on the ImageScan module being available.

ImageIn

ImageIn is the main program used to manipulate the other modules (see Fig. 9.1). It reads and stores scans in different file formats, offers several picture views and picture information and provides the interface to the Windows printer. It is not necessary to have a scanner to be able to use the functionality of ImageIn successfully. ImageIn works with file formats such as MSP, PCX, TIFF, PIC (MacPaint), GEM Paint, and EPS (PostScript). Scans or picture documents can be stored completely or partially. Picture size is controlled by means of rulers (either in inches or centimetres), and graphics can be viewed either at original size or in a reduced view. ImageIn can also provide data about the graphic being viewed (see Fig. 9.2). From ImageIn we can print scans, and vary the size of the image. The *Undo* function can be very useful, enabling any change made to pictures to be undone.

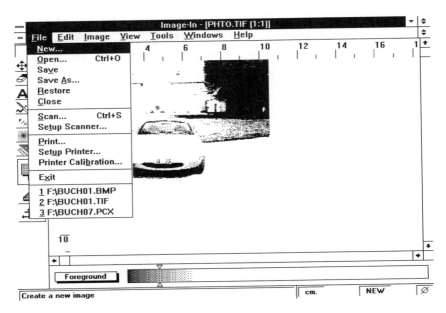

Figure 9.1 The main program ImageIn

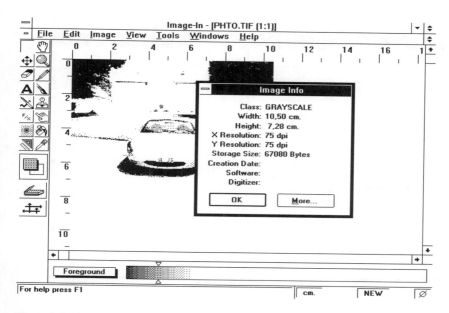

Figure 9.2 Picture information in ImageIn

ImageScan

ImageScan is the interface we use to control the scanner, via special scanner drivers that are delivered with the scanner. Because ImageScan is supplied independently, it uses the scanner drivers that are supplied with the scanner, and therefore can utilize any special features the scanner has. In ImageScan we can easily adjust scan parameters for brightness, contrast and resolution (see Fig. 9.3). Using ImageScan, we can work with line drawings, dithered and grey scale pictures, as long as the installed scanner is also capable. We can select a scan area to scan a document in a mixed line/dithered mode. The maximum size of a scanned picture is restricted only by disk capacity, and not by the amount of main memory that is available. This is an important advantage compared with many other scan programs. Depending on the scanner we are using with ImageScan, we can choose scan modes for line art (1-bit), halftones (1-bit—simulated grey tones) and grey steps (up to 8-bit and 256 real grey tones). Of course we can also adjust scan parameters for brightness, contrast, resolution, for parts of a picture, scaling and so on. Some scanners also offer an option for 'mixed digitizing' (one picture halftone, the other one line art). The scan parameters that have been selected can be stored for further use.

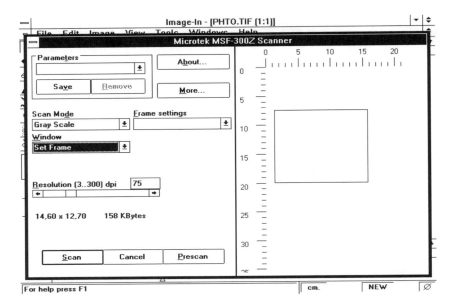

Figure 9.3 The ImageScan module

ImagePlus

We have already pointed out in earlier chapters the difference between real grey scale scans and those that only simulate grey scales using halftone patterns. ImagePlus can support up to 256 shades of grey that can be created using scanners from Microtek, Hewlett-Packard and Agfa, for example. Before buying this module we need to make sure we can use it to drive our scanner. ImagePlus provides a very flexible way to manipulate grey scale scans and may be considered to be a kind of 'digital darkroom'. Complete pictures or parts of them may be retouched, and filters used for special effects including line art transformation, edge enhancement, sharpening, blurring, softening, noise enhancement and noise reduction of a grey scale scan (see Fig. 9.4). ImagePlus offers a number of options for editing and optimizing the gradation of grey scale scans. We can change the gradation curve by entering a particular value or by defining the curve with the mouse in a special window (see Fig. 9.5). The ability to filter and adjust the gradation simplifies the scanning process enormously, and we do not have to make several scans, because the optimizing of the scan takes place during picture manipulation at a later stage. Because of this we can work off-line, the facilities of the scanner no longer being needed. To demonstrate the effects of picture manipulation with grey scale scans in ImagePlus, we have taken an image scanned quite normally with 256 grey steps (see Fig. 9.6) and applied different filters and

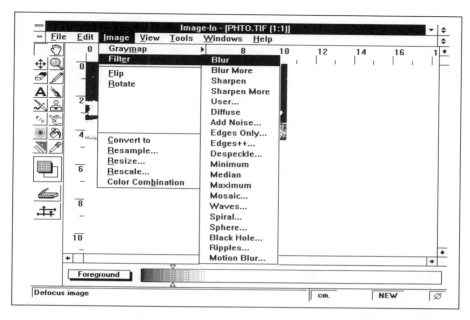

Figure 9.4 Filter effects for grey scale scans in ImagePlus

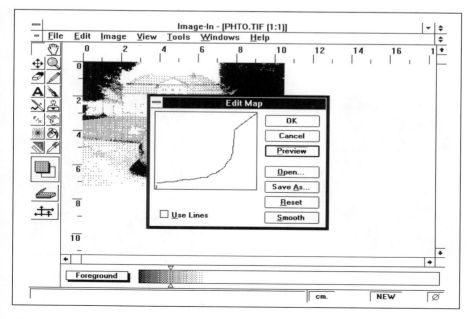

Figure 9.5 Editing of the gradation in ImagePlus

gradation curves. Figures 9.7 to 9.12 show the results. For a partial retouch of grey scale scans ImagePlus provides different brush shapes, spray densities and the choice between opaque and transparent. We can also cut, copy and insert parts of a scan. ImagePlus also gives us all the functions of the ImagePaint module, which is used to edit line art scans and dithered halftone scans.

Figure 9.6 Scan with 256 grey steps before editing

Figure 9.7 Using the line art filter (15%)

Figure 9.8 Sharpening with 100%

Figure 9.9 Softening with 100%

Figure 9.10 High contrast

Figure 9.11 Poster effect

Figure 9.12 Solarization

ImagePaint

If we do not need the amount of function in ImagePlus, we can use ImagePaint instead. This module offers editing and paint functions for line art scans or simulated halftone scans. We can also use many of the tools for grey scale scans. ImagePaint recognizes and automatically loads picture files in the following formats: GEM Paint, EyeStar, MacPaint, Microsoft Paint, PC PaintBrush, Windows Bitmap, Windows PM/Bitmap, PostScript (EPS), and TIFF. The amount of available memory determines how large an image we may work with, and it is possible to work in a strongly magnified mode (see Fig. 9.13). In addition to scan editing we can draw simple objects like rectangles, squares, ellipses, circles, polygons, lines and so on with different line weights and fill patterns. The program offers a Bézier curve control and the option to insert text into a scan, for which any of the fonts that are installed in Windows may be used. Figure 9.14 shows a scanned line art image—before and after editing in ImagePaint.

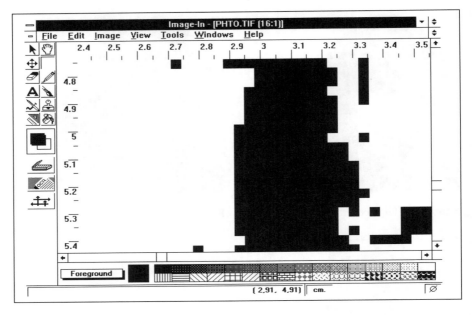

Figure 9.13 Line art in magnified mode in ImagePaint

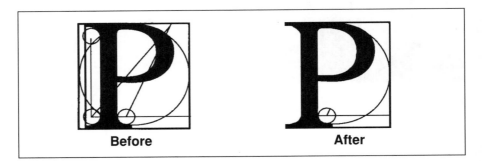

Figure 9.14 Line art before and after editing in ImagePaint

ImageRead

The ImageRead module translates scanned text documents in line art mode into ASCII files (see Fig. 9.15). Alternatively, we can read texts put into the Windows clipboard and transfer them from there to other Windows applications. Most text processing programs (such as word processors) can read ASCII files. ImageRead is an OCR program, and we will mention these only briefly in this book. ImageRead is quite easy to use, largely because the learning mode that is available during the character recognition offers a high amount of flexibility. ImageRead can recognize the complete IBM character set—in a range of fonts, font sizes and paragraph formats. The combination of ImageIn and ImageRead provides us with the means to edit pictures as well as texts from a particular document.

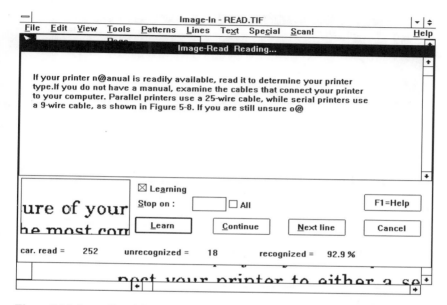

Figure 9.15 ImageRead for optical character recognition

ImageVect

ImageVect translates bitmap scans or parts of scans into vectorized graphical objects. These graphics can then be easily transferred into CAD/CAM programs such as AutoCAD, In-A-Vision, Designer and CorelDRAW for further editing. EPS format may be used to transfer images into DTP programs like PageMaker. The advantage of vectorized scans compared with normal bitmap scans is that they may be enlarged or reduced without loss of resolution; they are also usually smaller in the amount of disk space they use. Figures 9.16 to 9.18 show printouts of a bitmap scan that has been reduced in size by 50 per cent and enlarged by the same amount. It has not been edited, and it has been scanned in line art mode at a resolution of 100 dpi. We then vectorized the scan using ImageVect and printed the result at 50 per cent and 150 per cent of the original size. ImageVect can also be used to vectorize halftone scans, converting photographs into line art (Figs. 9.19a and b). For the best results a halftone picture should be scanned in line art mode with relatively high contrast settings, and then be vectorized (see Figs. 9.20a and b). There are other means to vectorize scanned images, besides ImageVect, for example, CorelDRAW with CorelTRACE, Streamline and Designer. We will describe these programs below.

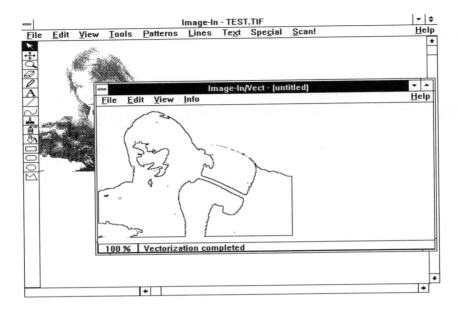

Figure 9.16 Vectorizing bitmap scans using ImageVect

Figure 9.17 Bitmap scan, line art mode, 100 dpi

Figure 9.18 Vectorized bitmap scan, line art mode, 100 dpi

Figure 9.19a Grey scale scan before vectorizing

Figure 9.19b Grey scale scan after vectorizing

Figure 9.20a Halftone image scanned in line art mode, before vectorizing

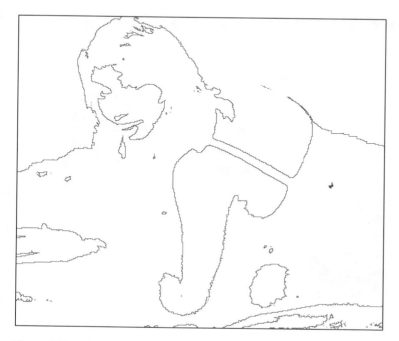

Figure 9.20b Halftone image scanned in line art mode, after vectorizing

Panorama

Panorama is a program that supplements ImageIn, and it helps us to organize graphic images. It enables us to organize, view and choose picture files that need not have been created using ImageIn—we can also work with graphics that have been created using other Windows or MS-DOS programs. The picture files in Panorama are organized by keywords and are identified by text descriptions. A graphic may be viewed on the screen on its own (main window) or in catalog view (Fig. 9.21). The search for a picture stored on disk is accomplished using a file name or a keyword that can be up to 16 characters long. For each graphic on the screen the user can see useful information such as the complete file name, date and time of creation, format, file size, keyword and a description that can be up to 255 characters long. For data exchange with other Windows programs the reduced version of the image may be used to transfer the entire graphic into the Windows clipboard—simply by clicking on the graphic. We can add picture files to the database or delete them, at any time, and keywords and descriptions can be changed. Updating of a database may be done either automatically or manually. Panorama works with the following graphic file formats:

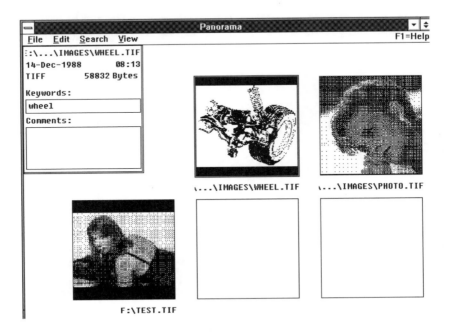

Figure 9.21 Catalog view of the Panorama picture database

TIFF	compressed or uncompressed
MSP	Microsoft Paint, versions 1 and 2
PCX	PC PaintBrush
PIC	Lotus 1-2-3, Symphony
DRW	Draw, In-A-Vision, Designer
MacPaint	

We can print lists of the picture files in the Panorama database. The printout gives us the file names as well as the directory paths, keywords and comments.

GRAY F/X

Gray F/X is another program we can use for bitmap editing on the IBM PC and compatibles. The program works only with TIFF format picture files, and it is not possible to import or export in other file formats. Gray F/X works in a totally different way from ImageIn and other programs that run under Microsoft Windows. Gray F/X is advertised as 'the first powerful and easy to handle electronic darkroom for the PC'. This may be an exaggeration, but it gives us an idea of the features of the program. Apart from providing numerous picture manipulation capabilities, it also provides drivers for a range of scanners and printers. For the purpose of printing images, it should be noted that PostScript printers can only be connected via a parallel port: this means that printers such as the Apple LaserWriter may not be used. We use Gray F/X in the same way as we use any other Windows program, with a mouse. The mouse driver (for example MOUSE.SYS or MOUSE.COM) has to be installed or activated separately from the operating system before we start the program. Figure 9.22 shows Gray F/X with multiple windows: two of these windows (or 'workspaces') already contain scanned grey scale images.

Gray F/X shows shades of grey very accurately on a VGA screen, and it is not possible to give an accurate printed representation in Fig. 9.23. In Gray F/X up to ten workspaces or scans can be worked on simultaneously. The program manages the necessary memory very intelligently; the user does not have to provide any information about memory—neither in the program itself nor during installation. Figure 9.23 shows the memory status after choosing the option AVAILABLE IMAGE MEMORY. Gray F/X uses any additional memory, expanded or extended, and shows the amount of memory used. If there is no extended or expanded memory available, Gray F/X uses the hard disk.

Gray F/X can show several editable pictures on the screen at the same time. The user can combine the pictures, cut certain parts of the pictures and insert them. The picture control, even for grey scale images, offers the capability to rotate and transform pictures and to change their size. Pictures may be enlarged by up to 1600 per cent—this enables the editing of even the tiniest details. If scanned images contain dirt or spots, we can easily get rid of them

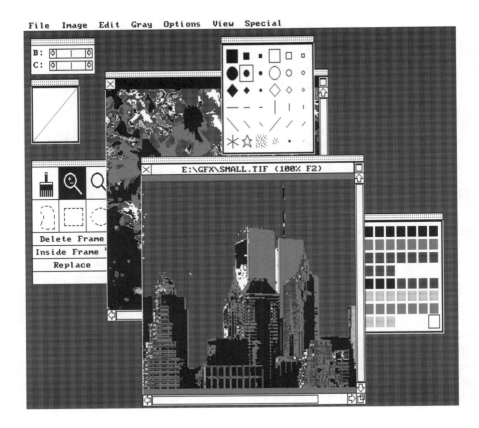

Figure 9.22 Gray F/X with two scanned grey scale images

by using the *Despeckle* function. By using a mask we can enhance or hide certain parts of a picture. For electronically retouching, Gray F/X offers an assortment of brushes and sprays (from hair-line to extra big). Scans can be enhanced using effects like filtering, reducing grey values, or inverting, and skilful use of an appropriate effect can enhance a very average photograph immensely. All the tools may be used in turn on an image, so that a scanned image can be changed in many ways. The *Undo* function is invaluable, if we change our mind about some alteration we have made, and want to revert to the previous version. There are numerous adjustments that may be made to brightness and contrast in Gray F/X. There are options to adjust existing grey tones via special histograms.

The following figures illustrate some of the most important functions of Gray F/X in manipulating pictures. For the examples, we started out with a black & white photograph that was scanned at a resolution of 75 dpi in grey scale mode using a Hewlett-Packard ScanJet IIp with 256 grey tones (see Fig. 9.24).

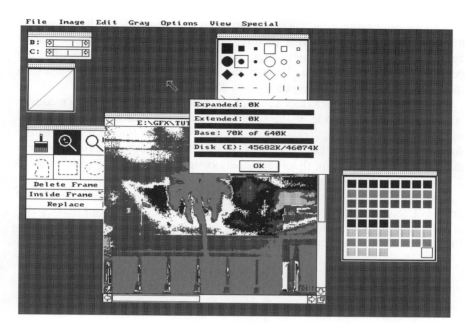

Figure 9.23 Intelligent memory organization in Gray F/X

Figure 9.24 The original grey scale scan used to demonstrate picture manipulation in Gray F/X

We can adjust the grey tones that exist in a scan. We can determine the number of grey steps, invert a picture, and equalize the grey values (linear, bell, exponential, logarthmic). Figures 9.25 to 9.27 show some of these possibilities. Grey maps that have been changed can always be set back to their default state by choosing the option *Reset Gray Map*.

The filters in Gray F/X can be used to different degrees (small, medium, large). When selecting one of the options, the filter chosen is directly applied to the image or a selected frame. The speed of a selected filter is related to the size of the image or the frame. As the filters work cumulatively, extreme effects may be obtained by using them repeatedly or in combination. Figures 9.28 to 9.32 show some of the filter effects.

The functions for retouching in Gray F/X enable us to edit single picture details. There are plenty of options to choose from so that we can make parts of an image lighter or darker, replace by certain grey values, blur, or blend. Figure 9.33 shows an example which uses some of these retouch functions.

Figure 9.25 Linear equalizing

Figure 9.26 Reducing the number of grey values to four

Figure 9.27 Inverting

Figure 9.28 Sharpen filter (strong)

Figure 9.29 Blur filter (strong)

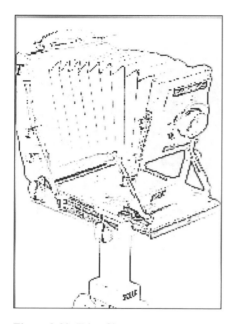

Figure 9.30 Edge filter

Figure 9.31 Spatial posterize filter (reduced pixel to 4 x 4)

Figure 9.32 Despeckle filter (strong) to remove unwanted 'noise' or patterns

Figure 9.33 Set apart and retouched scan

The editing functions offer the normal capabilities to cut, copy, and paste within a document. This can take place in the active workspace or in a new, empty one. Other controls for pasting or inserting include replacing or blending at different percentage levels. Pictures or parts of pictures can easily be transformed and mirrored. Figure 9.34 shows the result of using some of the editing functions in an exercise that was not intended to produce a professional result!

If we look more closely at the scanner and printer interfaces in Gray F/X we can see below, first of all, the range of scanners that can be driven directly from the program (Fig. 9.35):

Scanners available in Gray F/X

Canon IX-12
Canon IX-30F
Datacopy Jetreader 730, 730 GS, 830
DEST PC Scan 650
HP ScanJet IIp, ScanJet Plus
IBM 3117, 3119
Microtek MS-300A, MS-300C, MS-300G
Panasonic FX-RS 505, FX-RS 506
Princeton Graphics LS 300A
Ricoh IS-30, SS-30, RS-320
Sharp Color Scanner

Printers can only be connected to Gray F/X using the ports LPT1 or LPT2 or via DOS. The following printers can be used with Gray F/X:

Printers available for Gray F/X

Canon LBP-8A2
Epson GQ-3500, LQ-2500, MX-80
Fujitsu DL2400
Hewlett-Packard DeskJet, LaserJet+, PaintJet, ThinkJet
Intel Visual Edge
Kyocera F-1010 Compact Laser
Laser Master Cap Card, LX-6
NEC P5XL
Okidata LaserLine
Panasonic KX-P1524, KX-P4450
PostScript printer (parallel)
QMS PS 800 II, Quadram Quadlaser
Ricoh PC Laser 6000, Toshiba P351, P1351
XEROX 4045 Laser CP Model 50

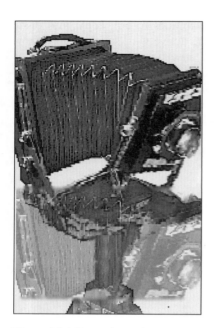

Figure 9.34 Scan after using some of the editing functions

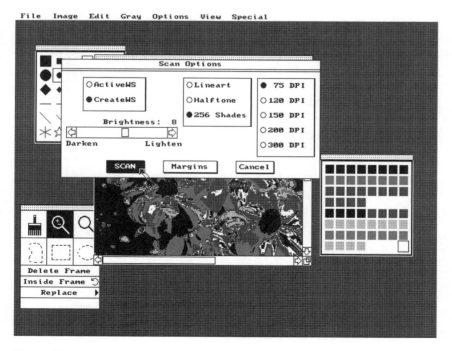

Figure 9.35 Scanner manipulation in Gray F/X

It is possible to print halftones from Gray F/X, via a special window that enables us to select different screens, depending on the printer (Fig. 9.36). The halftoning options are recommended for use with 300 dpi printers only.

It is not possible to discuss all of the features of Gray F/X in this book; we recommend that those who do a major amount of grey scale manipulation should consider purchasing this very functional program.

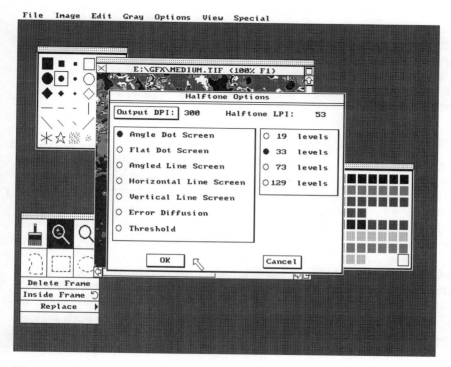

Figure 9.36 Selection of halftone options for printing

PICTURE PUBLISHER

Picture Publisher is another program that enables us to manipulate grey scale scans. The version we used was 3.0, running in the Microsoft Windows 3.1 environment. We were most impressed with Picture Publisher's context-sensitive help, which responded very effectively as we moved the mouse to different features on the screen, and pressed the F1 key.

Picture Publisher permits us to edit pictures of any size and resolution, independent of the memory configuration or the actual memory available. Input and output devices (scanners, frame grabbers and printers) can be used directly from Picture Publisher, and it is possible to save calibrations made for using these devices, so that they can be used again, under identical conditions. Picture Publisher reads grey scale and line art pictures in TIFF format, and both uncompressed and compressed files that have been saved in TIFF 5.0 format. Export formats may be found in Fig. 9.54.

Scanning

Many of the most popular scanners may be driven directly from Picture Publisher, so that in many cases where the program is to be used for editing pictures, no additional scan program is necessary (Fig. 9.37). Scanners are driven via an integrated scanner interface. These scanners include models produced by Canon, Epson, Hewlett-Packard, Microtek, Panasonic, and Siemens.

Depending on the scanner, we can choose a variety of settings for resolution, mode (grey scale or line art, scaling and so on). After prescanning a picture, we can select the part required before making a final scan with whatever settings we want. As soon as the scanning process is complete, we will see the screen shown in Figure 9.38. We see the same screen when we load an existing picture file.

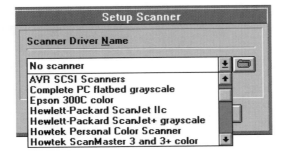

Figure 9.37 Selecting a scanner in Picture Publisher

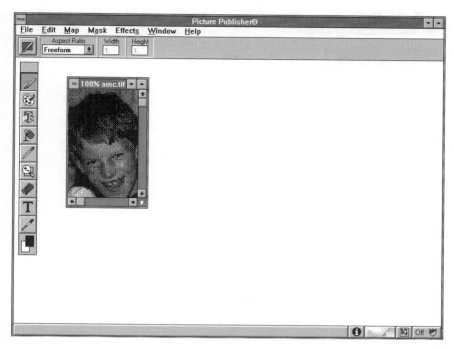

Figure 9.38 Picture Publisher after scanning or loading a picture file

Editing images

Picture Publisher uses a combination of on-screen icons (Fig. 9.39), and menu drop-downs to enable us to select what we want to do to the image on which we are working.

Strengths of Picture Publisher

Clearly, with some of the sophisticated software that comes with scanners these days, we get an immense amount of functions. Applications such as Picture Publisher are needed where our scanner may be older, and was not initially supplied with image processing software, or where we need a function that does not exist in our own image processing software. Most software, for instance, will permit us to invert an image so that it becomes a negative. Most will enable us to rotate an image by 90 degrees. Picture Publisher takes this manipulation a stage further and lets us rotate images at arbitrary angles, either clockwise, or anti-clockwise (Figs 9.40 to 9.43).

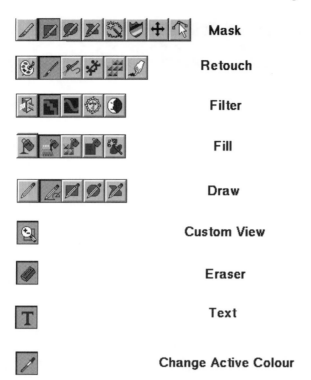

Mask

Retouch

Filter

Fill

Draw

Custom View

Eraser

Text

Change Active Colour

Figure 9.39 Tool icons for Picture Publisher

Figure 9.40 Image, scanned normally

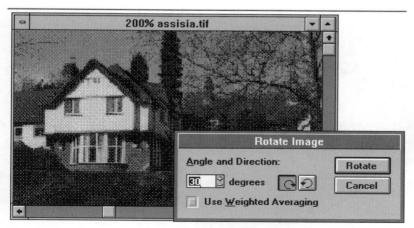

Figure 9.41 Selecting an angle for the rotation—30 degrees

Figure 9.42 The image rotated—on screen

Figure 9.43 The rotated image, printed

Once we have scanned an image, we may not always want to print it, or use it, as an identical copy of the original. We may want to change it out of all recognition or to use it as the basis for some special effect. Picture Publisher provides us with a wide range of filters to effect such manipulation.

One of the advantages of a powerful graphical utility such as Picture Publisher is that we can compare different effects on the screen as we work, without needing to print first. Figure 9.44 shows nine different effects—all very different—in the Picture Publisher workspace.

From the examples in Fig. 9.45, we can see that Picture Publisher contains a wide-ranging capability for applying filters to images or parts of images.

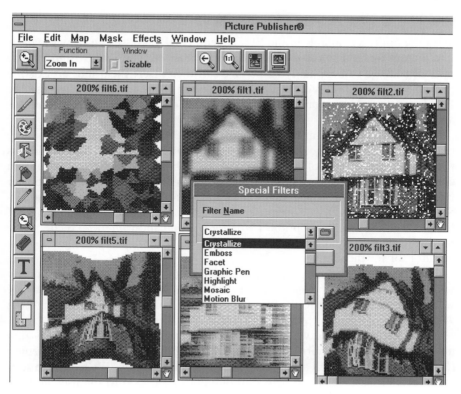

Figure 9.44 Comparing the effects of different filters on the screen

Figure 9.45 Top row, left to right: original image, smooth edge filter, 10% splatter, twirl 45 degrees. Bottom row: wind, 3-dimensional, crystallize.

Accessing other image processing capabilities

With the options available in the EDIT menu (Fig. 9.46) we can copy the parts
of a picture we have selected into one of a number of clipboards (using the *Cut*
option) that we can explicitly name. We can then insert them into an existing
picture (using *Paste*). We can *Delete* a masked area. In general, we use the
EDIT menu only in combination with some of the other features of this
application.

We use the MAP menu (Fig. 9.47) to make either global changes to a grey
scale image, or to change a masked area. By changing contrast and brightness
the overall impression of a picture may be enhanced. The *Posterize* option
enables us to reduce the amount of grey steps to create what is called a poster
effect. We can then adjust highlights and shadows (*Modify Color Maps*).
Selecting the *Quartertones* option enables us to adjust the dynamic range of
the picture. The final two options (*Monitor Gamma* and *Calibrate*) relate to the
hardware of our computer system, and help to calibrate the images for the
monitor, or for the scanner or printer.

Once we have defined a mask, using one of the masking tools, or by
selecting *Mask Entire Image* from the MASK menu (Fig. 9.48), we can apply
whatever effect we want to the masked area. We can also ensure that anything
we paste in will blend with the rest of the image (*Blend Mask*).

The function provided via the EFFECTS menu is self-explanatory for sizing
images, rotating and mirroring them, inverting, smoothing and sharpening
(Fig. 9.49). We have seen all of these elsewhere in this book. We have also seen
the effects of edge detection before, though not described to the user in quite
the same way (Fig. 9.50).

Figure 9.46 The EDIT menu

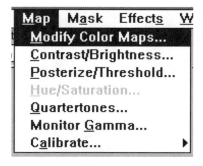

Figure 9.47 The MAP menu

Figure 9.48 The MASK menu

Figure 9.49 The EFFECTS menu

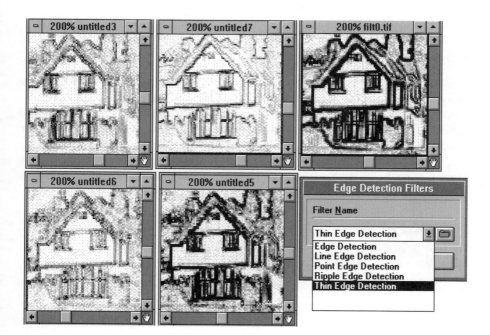

Figure 9.50 EDGE DETECTION selection list and effects

Retouch

The *Retouch* icons enable us to optimize a picture using various brushes with different pressure settings and paint styles. We can see the current retouch settings in the ribbon bar at the top of the Picture Publisher window (Fig. 9.51). We can spray grey tones on to parts of an image, copy textures into other parts, or duplicate parts by pixel into other areas.

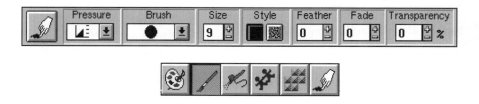

Figure 9.51 The *Retouch* icons

Mask

The *Mask* icons (Fig. 9.52) are used to outline and activate certain parts of an image so they can be treated separately. The user can decide at any time whether the changes should take place within, or outside the outline of the mask. Masks may be created in rectangular, square, elliptical or circular shapes. The *Freehand* option enables us to draw a mask in any shape we want. The mask outline that we have drawn may be erased using the backspace key. Once a mask has been created, it will remain in the picture, no matter how it is sized.

Figure 9.52 The *Mask* icons

View

The *View* icon ( ) allows us to see the grey scale image at a range of enlargements when we are working in editing mode.

Effects

The EFFECTS menu (Fig. 9.49) contains functions to change the resolution, dimensions and orientation of a picture. Global filter effects may be applied to an image. Since a change in size often means a reduction in quality, Picture Publisher uses a special optimizing algorithm that can be activated with the *Use Smart Sizing* option.

We distorted the scan in Fig. 9.53 by reducing the width by 50 per cent. In this figure we can see both the original image and the result of our manipulations, together with the *Size* window, with the option Allow Size Distortions selected. Normally, we will wish to maintain the aspect ratio of our scanned image. There will be times, however, when for special purposes we will want to vary it. The manipulation took a matter of seconds, and the final product can be further edited in Picture Publisher if required.

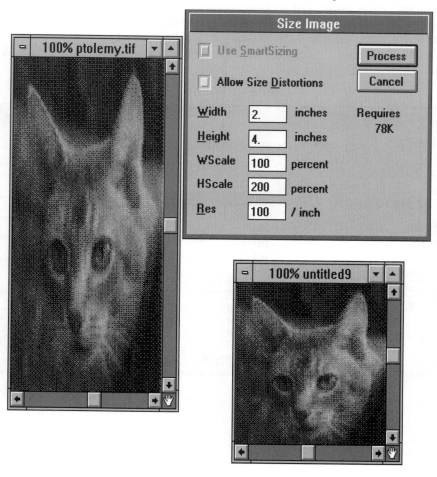

Figure 9.53 The *Size* option in the EFFECTS menu

Formats

In Picture Publisher we can save the pictures we have in a variety of common formats. These include TIFF, Windows Bitmap (BMP), PC PaintBrush (PCX) and EPS (Fig. 9.54).

We can also choose to save our image as grey scale, line art and scattered patterns (Fig. 9.55). If we select the *Line Art* option, a grey scale scan will be saved as line art, using only with black and white information. With the *Scattered* option the image is saved in a unique style as dithered image with enhanced edge detail. If we only want to use the image later on in Picture Publisher (and not in another program such as PageMaker), we can use the option *Compress* to compress an image in order to save storage space. The compression is compatible with the latest versions of the TIFF format, TIFF 5.0, but applications such as Aldus PageMaker will not accept such a file. To summarize the features of Picture Publisher; this program is a very powerful means of manipulating images. It has interfaces to many scanners, frame grabbers, printers and high resolution video cards. The program is very intuitive in use, and the editing mode is particularly powerful, as a wide range of manipulating options is available.

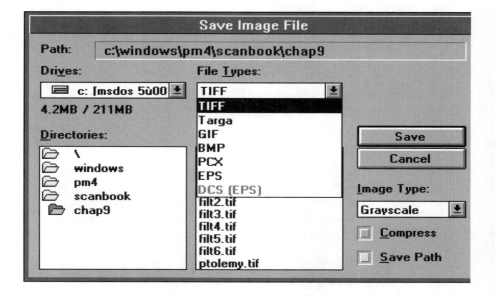

Figure 9.54 Options for saving files in Picture Publisher

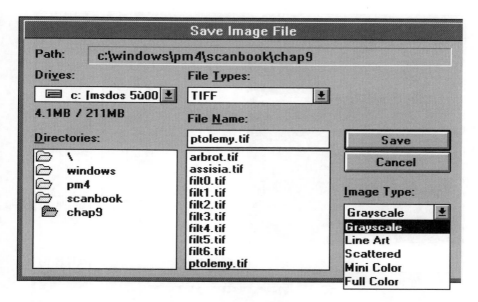

Figure 9.55 Options for image types in Picture Publisher

PHOTOSTYLER

PhotoStyler is yet another program that enables us to accomplish pretty much the same types of picture manipulation. Whether we find its user interface to be more intuitive than that of any of the other programs is a very personal thing. It certainly gives us a mass of tools to display on the screen as we go about the business of image-bending (see Fig. 9.56).

Converting image types

Once we have opened a scanned image file, we can convert it using the IMAGE menu (Fig. 9.57). This provides us with many of the effects we have seen before in the other applications, with a number of quirky, but interesting additions. Whether we will ever need to use these effects depends, of course, on what we want to scan for in the first place! (Fig. 9.58 to 9.65)

COMPARING PROGRAMS FOR IMAGE PROCESSING

For normal demands, ImageIn and EyeStar II are sufficient for scanning and picture editing, as we can work quite easily with grey scales as well as with line art, where EyeStar offers additional colour capabilities. If we need a wide range of functions for grey scale manipulation (or retouching), then Gray F/X

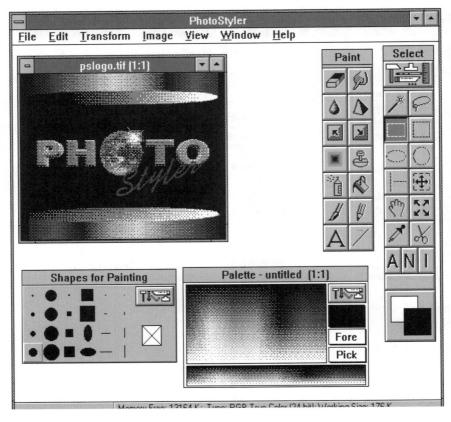

Figure 9.56 PhotoStyler main screen and tools

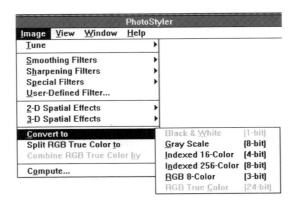

Figure 9.57 PhotoStyler—converting images

may be the best choice. Gray F/X gives us the advantage of being able to work simultaneously in up to ten workspaces. This is important when we want to combine parts of different images to make a new picture. ImageStudio claims to be a professional program but, although it is very expensive, it is necessary to have a very well-configured system in order to be able to view grey scale images on the screen. The other programs can display grey scales on standard PC configurations and this is a distinct advantage, as only when we can see exactly what we are working on can we realistically attempt to manipulate grey scale images.

Figure 9.58 The 3-D sphere effect on a selected area

Figure 9.59 Using 3-D pinch

Figure 9.60 The 3-D cylinder

Figure 9.61 2-D ripple

Figure 9.62 PhotoStyler: trace
contour and select negative

Figure 9.63 PhotoStyler 2-D
wave effect on selected area

Figure 9.64 2-D whirlpool effect

Figure 9.65 Effect of the special filter 'minimum'

CORELDRAW

The graphic design program CorelDRAW runs under Microsoft Windows. It is an ideal supplement to desktop publishing and scanning programs. In CorelDRAW, we can use more than 100 different fonts in any size and apply very interesting effects to them in the same way as the effects can be used with any graphical objects. We can create drawings using Bézier curves and fill objects with colours and special PostScript patterns. Drawings and single objects can be moved, stretched, rotated, mirrored, slanted and copied. This wealth of options makes CorelDRAW ideal for graphic design, making illustrations, creating logos and advertising headlines as well as creating presentations using graphics, charts and diagrams. Output can be sent to any printer, plotter or slidemaker for which Windows has a driver. Figure 9.66 shows a drawing on the CorelDRAW screen.

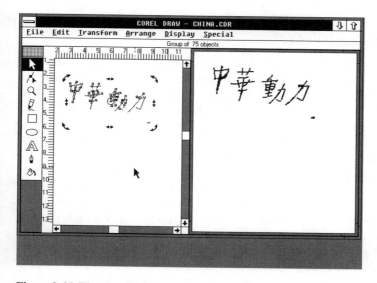

Figure 9.66 The CorelDRAW screen with a drawing

With CorelDRAW we can import and export the following picture and graphics files:

Import   **Export**

CorelDRAW	CDR	*as Import, plus*	
CorelTRACE	EPS	Encapsulated Postscript	EPS
PC PaintBrush	PCX	Matrix SCODL	SCD
Windows Bitmap	BMP	Targa Bitmap	TGA
Windows Metafile	WMF	WordPerfect graphic	WPG
AutoCAD	DXF	Adobe Type 1 font	PFB
Compuserve Bitmap	GIF		
Computer Graphics Metafile	CGM		
GEM	GEM		
H-P Plotter HPGL	PLT		
IBM PIF	PIF		
Illustrator 88	AI, EPS		
Lotus PIC	PIC		
Mac (PICT)	PCT		
TIFF 5.0 Bitmap	TIF		
text	TXT		

For the scanner user, one of the most useful features of CorelDRAW is the ability to import bitmaps in TIFF and PCX formats. These formats are, of course, those used by scanning programs, and we can use CorelDRAW to manipulate them. A companion program CorelTRACE enables us to trace bitmaps. We describe these options below.

Scanning and CorelDRAW

Suppose we want to create a business logo, but we are not confident of drawing it by hand. We have a scanner and a scan program available to help us to use the photograph in Fig. 9.67 as a basis for our logo. We scan the photograph for the logo in line art mode at 100 dpi resolution (we actually recommend 300 dpi) and store it in a file format that can be imported in CorelDRAW. For this example we used TIFF format (see Fig. 9.68). When we have scanned the picture, we exit the scan program and load CorelTRACE, which comes as part of the CorelDRAW package (from Version 1.2 onwards). We use CorelTRACE to vectorize bitmap scans, though there is a trace option available within CorelDRAW itself that is very much less sophisticated, but nevertheless useful for small amounts of tracing. The disadvantage of bitmap scans is that, once an image has been scanned, the resolution is fixed, and no matter how high the resolution of the printer that is to be used, the resolution of the print will remain as it was scanned. If we enlarge a scanned image and print it on a high resolution printer, there will be a noticeable degradation of the image (see Fig. 9.69).

Figure 9.67 The original to be used in the logo

Figure 9.68 Bitmap scan as line art (TIFF format) with a resolution of 100 dpi

Figure 9.69 A bitmap scan (resolution 100 dpi) printed with 200% enlargement

CorelTRACE solves the problem of outputting enlarged bitmap scans by translating them into vectorized images that can always be printed at the resolution of any printer that is being used, no matter how high its resolution. We can enlarge, reduce, or even rotate vectorized pictures without any quality loss. A further advantage of vectorized images compared with bitmap images is that they need less disk space for storage and far less printing time. After vectorizing we can transfer these pictures directly into a desktop publishing program or edit them further in CorelDRAW. If we return to our simple example with the logo, we use CorelTRACE to vectorize the photograph we have scanned. Before we do this we can define the pixel resolution, line width, and define whether a centre line or the outer edges should be traced. Frequently we can use the default values in CorelTRACE. After vectorizing the screen should look like Fig. 9.70. A vectorized scan is stored in Corel-TRACE either in AI (Adobe Illustrator) or EPS (Encapsulated PostScript) formats. Both formats are supported by CorelDRAW and most other drawing programs, as well as desktop publishing programs like PageMaker and Ventura Publisher. Once the vectorized scan is in one of these transport formats, we can transfer it directly into a desktop publishing program or manipulate it further in a drawing application. In this example, we imported the logo into CorelDRAW and used some of the program features, such as the filling of closed objects, rotating and combining with text. Besides the drawing options (rotating, skewing, stretching, mirroring, duplicating, and so on) CorelDRAW offers many different fonts that can be used at any size as they come, or modified. As Fig. 9.71 shows, we can also combine the image with text. We exported the logo as an EPS file so that we could print it out at a later stage from PageMaker. Figure 9.72 shows the result, which was printed at a resolution of 300 dpi.

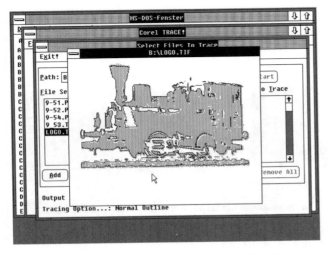

Figure 9.70 Vectorized bitmap scan in CorelTRACE

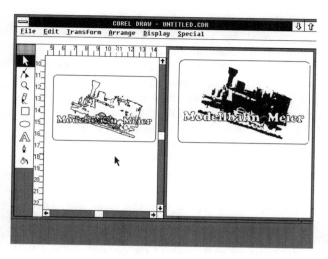

Figure 9.71 The CorelDRAW screen during manipulation of the logo

Figure 9.72 The finished logo—printed at 300 dpi

DESIGNER

Designer 3.0 is a graphics program similar to CorelDRAW. It also runs under Windows and is well known to those who work professionally with computer graphics and drawings. Figure 9.73 shows a typical Designer screen with an illustration. Designer 3.0 works with vectorized drawings in a similar way to CAD programs, such as AutoCAD; that is to say, it uses the technique of layers. A complex drawing is divided into elements that are then combined (on top of each other) to form an image. This way of working gives the user additional control and minimizes confusion of detail in large drawings. Drawing options are supplemented by powerful text element processing capabilities. The program offers more than 40 fonts that can be scaled up to 144 point (2 inches or 5 centimetres high). When the program is being used with scanned images, it may be used to trace bitmaps at a user-defined level of accuracy. Depending on the level selected, even the smallest of details can be traced and the outline smoothed. Unlike CorelDRAW, Designer 3.0 allows the tracing of coloured TIFF files, although this requires a system configured with expanded memory (LIM 4.0). Figure 9.74 shows a line art scan (scanned at 75 dpi) that is imported in Designer 3.0. During the tracing process this bitmap is constantly visible on the screen; when checking the vectorized drawing the bitmap can be hidden.

The vectorizing feature in Designer 3.0 uses a layering technique, and the final result is the line art drawing as one layer and the filled area as another. Both layers can be edited using the options of Designer 3.0. We can, for example, change the line widths, the fill patterns and the colours. Figure 9.75 shows the result of the vectorizing process.

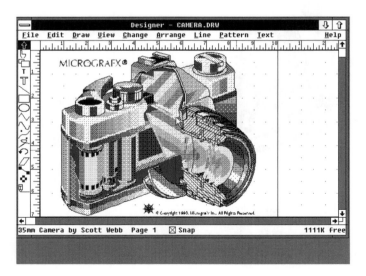

Figure 9.73 Designer screen with an illustration

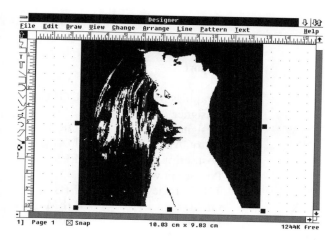

Figure 9.74 Designer 3.0 screen with imported scan

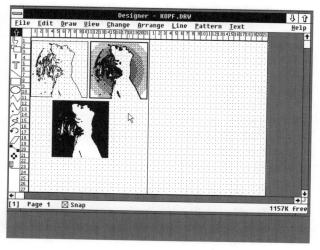

Figure 9.75 Vectorized scan in Designer 3.0

Designer 3.0 can import and export picture or graphics files in the following formats:

Import		Export	
Graphics Metafile	CGM	Graphics Metafile	CGM
Designer, Draw Plus		Designer 1.x, 2.0,	
Graph Plus	DRW	Designer 3.0	DRW
AutoCAD	DXF	AutoCAD	DXF
GEM	GEM	PostScript	EPS
Graph Plus	GRF	PostScript without	
MAC (PICT)	PCT	TIFF preview	EPS
PC PaintBrush	PCX	GEM	GEM
Windows Draw	PIC	HPGL	HP
In-A-Vision	PIC	MAC (PICT)	PCT
Windows Graph	PIC	PC PaintBrush	PCX
TIFF	TIF	Windows Draw	PIC
ANSII text format	TXT	In-A-Vision	PIC
Windows Metafile	WMF	Windows Graph	PIC
		TIFF	TIF
		Windows Metafile	WMF

The export capabilities of Designer 3.0 offer a variety of options for the TIFF format:

Option	Colours
Monochrome	Black and white
4-bit grey scale	16 grey steps
8-bit grey scale	256 grey steps
24-bit RGB colour	16.7 million colours
8 fixed colours	8 Windows colours
16 fixed colours	8 Windows colours, additionally each colour a bit darker
256 fixed colours	256 colours, in accordance with the IBM 8514 graphics card
Device colours	By monitor or printer supported colours

We found more problems exporting files from Designer 3.0 than we did with CorelDRAW. We believe this might have been due to the early version of the program that we used. Designer 3.0 is a professional illustration and drawing program, and the vectorizing options may be too comprehensive for the user who simply wants an additional program to edit scanned images. If we do not intend to become a professional graphic designer in our use of a scanner, we believe that it would be a more effective option to use the combination of

CorelDRAW and CorelTRACE. To give an impression of the many capabilities of Designer 3.0 we show in Fig. 9.76 an original and in Fig. 9.77 the version we have produced using Designer as a manipulation tool. Unfortunately, we cannot show the difference between the black and white original and the resulting coloured illustration.

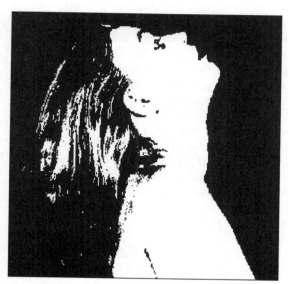

Figure 9.76 Black & white original, scanned in line art mode for editing in Designer 3.0

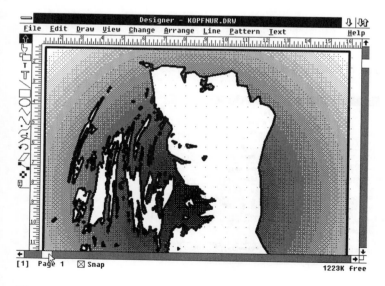

Figure 9.77 The manipulated illustration in Designer 3.0

ADOBE STREAMLINE

Adobe Streamline is only a tracing program. Like ImageVect and Corel-
TRACE, it runs under Microsoft Windows and can only be used to change
bitmap scans into vectorized images. Unlike the other programs, Adobe
Streamline can only work with true line art scans (1-bit data depth)—but it
does so in a very impressive way. Figure 9.78 shows the window with different
options for converting a scanned bitmap. Apart from choosing the conversion
method and the parameters we can also determine the lines to be traced, invert
a bitmap scan, or convert only a certain area of the scan into a vectorized
drawing. Most of the time the default values are sufficient.

In the next step we select the output format for the converted bitmap scan
(Fig. 9.79). Besides the formats for Adobe Illustrator (AI and EPS) we can
choose the CorelDRAW import format AI and the Designer format DRW.
The original bitmap (it determines the fill pattern) can be saved separately and
not as part of the vectorized conversion. If we want to edit the vectorized
drawing later on, we can work on this template on its own and then combine it
with the drawing. Figure 9.80 shows an example of this application that has
been created in CorelDRAW.

When the output options have been selected, we choose the bitmap file that
has to be converted (Fig. 9.81). The file can be in a TIFF, PCX, or PNT
format, though only with a data depth of 1 bit for black and white information. Having chosen the file name, we then have to confirm what we are
going to call the file to be saved (Fig. 9.82). Adobe Streamline suggests the
name of the input file, to which it appends the extension of the output file. By
clicking in the field CONVERT IMAGE(S) the conversion process is started.

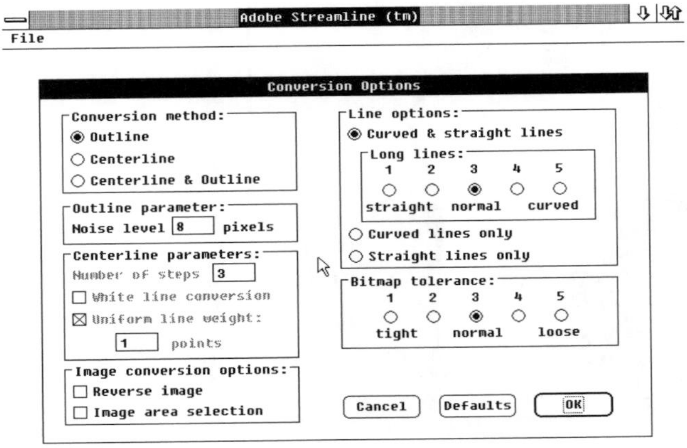

Figure 9.78 Adobe Streamline's conversion options

```
  ⊐ ▬▬▬▬▬▬▬▬▬▬▬Adobe Streamline (tm)▬▬▬▬▬▬▬▬▬▬▬     ⬇ 🔁
  File
```

```
┌──────────────────────────────────────────────────────────────┐
│                         File Setup                             │
│  ┌Art format:──────────────────┐  ┌Template format:──────┐     │
│  │ ⦿ Adobe Illustrator® (.AI)  │  │   ⦿ None              │     │
│  │ ○ Adobe Illustrator® (.EPS) │  │   ○ .PCX             │     │
│  │ ○ COREL DRAW!® (.AI)        │  │   ○ .PNT             │     │
│  │ ○ Micrografx Designer® (.DRW)│ │   ○ .TIF             │     │
│  └──────────────────────────────┘  └──────────────────────┘     │
│                                                                │
│  ┌Filename conflict:─┐  ┌Destination for converted images:─┐   │
│  │ ⦿ Always prompt   │  │   ⦿ Specified directory          │   │
│  │ ○ Never replace   │  │   ○ Original directory           │   │
│  │ ○ Always replace  │  └──────────────────────────────────┘   │
│  └───────────────────┘                                         │
│                                                                │
│               (    OK    )        ( Cancel )                   │
└──────────────────────────────────────────────────────────────┘
```

Figure 9.79 Output formats in Adobe Streamline

Figure 9.80 In CorelDRAW combined fill pattern for template and vectorized drawing

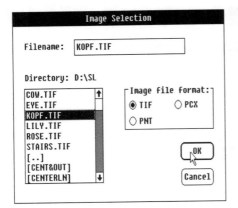

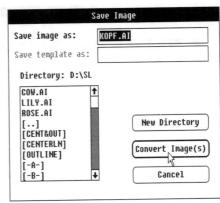

Figure 9.81 Selecting an input file   **Figure 9.82** Selecting an output file

The original bitmap scan appears on the screen (Fig. 9.83), and after a very short time (for this example, less than 30 seconds) the bitmap is converted to a vectorized drawing (Fig. 9.84).

After leaving Adobe Streamline we can import the converted bitmap scan directly into a desktop publishing program and print it. We can also import the Streamline output file (the vectorized bitmap scan) into a graphics program like CorelDRAW for further editing (Fig. 9.85). By changing the line width and the fill patterns we have a simple but effective means to manipulate a drawing (Fig. 9.86).

Figure 9.83 The bitmap scan before conversion in Adobe Streamline

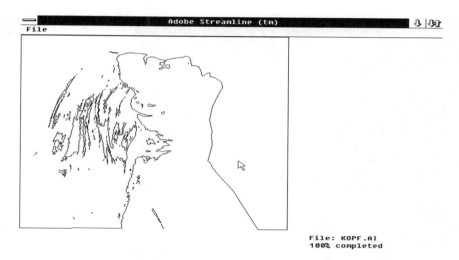

Figure 9.84 The vectorized bitmap scan in Adobe Streamline

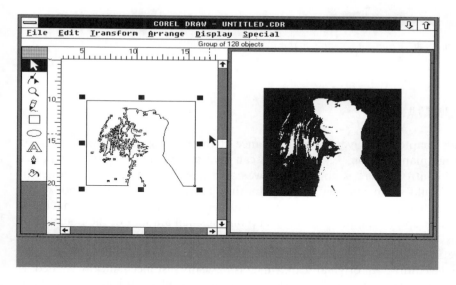

Figure 9.85 An Adobe Streamline conversion, imported into CorelDRAW

Adobe Streamline impressed us right from the start. The speed of converting and saving in a desired output format is far faster than the trace functions within CorelDRAW or Designer. It is possible to attain the same speed, however, using the stand-alone CorelTRACE program that comes with CorelDRAW.

Figure 9.86 Result of a simple manipulation in CorelDRAW

SUMMARY

This chapter has given us a first glance at the most important programs for image manipulation. Each of them can give a substantial amount of power and control. There is no way that we could describe each program in depth, nor could we mention every program that is available, within this book. We hope that we may have given the reader the understanding that having a scanner and a scan program opens up only a small amount of the potential of this 'art'. The many picture manipulation and graphics programs available for the PC offer every ambitious scanner user enormous possibilities for creative image editing. An original scan, which is basically just mechanically created, is only the first step along the road to a finished product. When we have been able to take advantage of all the different options and functions available in the graphics and image processing programs we can achieve results that are really worthwhile. The scans we show in Chapter 10 may give you some ideas for your own own creative work. Colour can lend another dimension, and a number of programs already offer this capability. Unfortunately, it is then necessary to invest quite heavily in additional hardware, such as a colour scanner and a colour printer.

Chapter 10
Gallery

Describing some of the features of image manipulation programs can only be a partial introduction to the subject unless it is also possible to see some examples of work that has been done using the programs. In this last chapter we would like to present examples that show the possibilities of the programs introduced in Chapter 9. All of the images were scanned using a flatbed scanner with 256 grey steps. We have restricted our 'gallery' to results produced using the most common output device, the laser printer. It is, of course, possible to generate results that include colour, or that use resolutions up to 2540 dpi if the appropriate means of output is available.

The examples in this chapter have been edited as TIFF or EPS files and then transferred into PageMaker. The prints were made with a 300 dpi laser printer.

Figure 10.1 'Bluebells'
This image was scanned at 256 grey levels on a Hewlett-Packard ScanJet IIp and imported to PhotoFinish, where it was edited (left), using *Edge Detect*, then contrast was adjusted (8485 bytes). The second picture shows the same image, but with the emboss option applied (23 152 bytes).

Figure 10.2 'Jan'
This image was scanned as a diffused halftone on a Hewlett-Packard ScanJet IIp. It was then imported to PhotoFinish, zoomed, then the image on the PC screen was saved using Collage (168 072 bytes).

Figure 10.3 'Cat on a hot sand beach'
We scanned a colour photograph of an Antiguan beach with 256 shades of grey on a
Hewlett-Packard ScanJet IIp. The result was imported to PhotoFinish, where we first
selected the sky, then inverted its colours. We applied blurring where we needed to
disguise joins. We copied and pasted part of another picture (a cat's head) and pasted
it onto the beach picture, and again used blurring to disguise joins (125 598 bytes).

The original beach picture (11 606 bytes)

Figure 10.4 'Bärbel'
We scanned the original for this picture with 256 grey steps, using Gray F/X, and used
that program to produce the required effect. File size: 161 662 bytes)

Figure 10.5 In this case a landscape photograph was scanned as if it were a line art image. File size: 10 568 bytes.

Figure 10.6 Here we scanned the same picture, this time using 256 grey steps. File size: 82 926 bytes.

Figure 10.7 Two overlapping colour slides were scanned with 256 grey steps to provide the basis for this picture. A photograph was then scanned, also with 256 grey steps. This picture was cut, rotated, then pasted into the original scan of the slides. All manipulation was made using Gray F/X. File size: 133 386 bytes.

Figure 10.8 A flatbed scanner, like a photocopier, can process objects other than photographs. In this case, we placed a stopwatch directly on the scanner and scanned it with 256 grey steps. Copies of the resulting image were made and these were reduced and distorted within Gray F/X before being overlaid on the original image—also within that program. File size: 148062 bytes.

Figure 10.9 Another object was placed directly on the scanner and scanned with 256 grey steps. This time it was enhanced within Eyestar II. File size: 55616 bytes.

Figure 10.10 A simple scan of a photograph—scanned with 256 grey steps.
File size: 95 642 bytes.

The original image has been retouched within Gray F/X. File size: 95 362 bytes.

Figure 10.12 We scanned a photograph of the exterior of a shop for Fig. 10.12 (file size: 33388 bytes).We then vectorized the image using Adobe Streamline (file size: 73432 bytes). Finally we imported the vectorized image into CorelDRAW and edited it further. File size 236310 bytes (EPS format).

Glossary

Adapter card   A circuit card, installed in the computer to provide additional features.

ASCII   (American Standard Code for Information Interchange.) In simple terms of text files, an ASCII file contains text characters only with no formatting such as bold, italic and so on.

AT-bus   Also known as ISA (Industry Standard Architecture) bus. Input/output architecture originally introduced with the IBM PC-AT. The most common bus in use on IBM PCs and compatible systems.

Automatic sheet feeder   An optional feature to enable automatic feeding of originals into the scanner.

Basic I/O address   Location in memory for a peripheral device.

Bitmap   A picture or graphic made up of dots or pixels rather than of objects or graphic instructions. A file that contains a picture in bitmap form contains information that describes each single point of the picture.

BMP   Windows 3 Bitmap file format.

CCD sensor   CCDs (charge coupled devices) are optoelectronic elements that record light as a charge in a condenser.

DCA   (Document Content Addressable.) A DCA file contains text and retains formatting such as bold, italic and so on.

Device driver   Program that allows the operating system to communicate with another device (printer, scanner, etc.). Under DOS these programs are usually activated via the CONFIG.SYS file.

Dither mode   Conversion of grey scale scans into halftone scans with a depth of a bit.

DMA (Direct Memory Access.) Should be used to enhance the scanning speed, if supported by the scanner driver. In case of clashes with other hardware devices, this option has to be deactivated.

DPI (Dots per inch.) Measurement unit for the resolution of a scanned picture.

Driver see *Device driver*.

EISA Extended Industry Standard Architecture. High bandwidth input/output bus introduced to give the benefits provided by the MCA bus, but also to be able to accommodate AT-bus interface cards. Not available for PC systems produced by IBM.

EPS (Encapsulated PostScript.) File format used by most page layout programs. The EPS format was developed by Adobe as an output format for PostScript printers. See *PostScript*.

Format Way of saving a picture format in a file. Each format needs a special algorithmn that allows the conversion of a picture in file form or a file in a picture.

Grey scale In grey scale pictures each point has a value that determines the intensitiy of its grey step. Black and white photographs, for instance, can contain a great number of grey steps. When editing with a computer, the number of grey steps is usually reduced (to 2, 16, 64 or 256).

Grey shades The intensity of a grey value in a picture. During the scanning process, the grey scale for each area is determined and then transferred to the computer.

Halftone A way to give the illusion of continuous tone. Black areas are represented by large black points, light areas by small black points.

HP-GL Abbreviation for Hewlett-Packard Graphics Language, a command language for plotters or for data interchange.

LED (Light Emitting Diode.) Used to sense brightness value for points on an original being scanned.

LIM (Lotus-Intel-Microsoft.) A joint standard for expanded memory cards that increase the amount of memory available for DOS applications. Not used by Microsoft Window 3.

MCA (MicroChannel Architecture.) High band width (fast!) input/output bus first introduced with the IBM PS/2 range of computers. Cannot accommodate AT-bus interface cards.

OCR   (Optical Character Recognition.) Technology to recognize characters during the scanning of a picture and to make them capable of being imported to a word processor.

Pattern   Simulation of grey values in images. A pattern simulates different grey shades by organizing identical points (point patterns) or lines (line patterns) in different densities The points or lines have always the same size. See *Halftone*.

Pattern angle   Determines the angle in which the pattern points or pattern lines are organized. The best results are usually obtained with an angle of 45 degrees.

PCX   File format originally developed for the PC PaintBrush painting program. Now used by many other programs as an interchange format.

Picture   Document containing only graphical elements.

Picture sensor   Optical sensor that converts light to electrical signals.

Pixel   Abbreviation for pixel elements as measurement unit. The size of a pixel differs from one device to another and consists of single, square units of the same size. Used for monitors, scanners and printers. 300 pixels per inch is an absolute measurement.

PostScript   A page description language designed by Adobe to be a standard output format. The language describes graphics not per point but vectorized. See *EPS*.

Printer point   Smallest element that can be printed. See *Printer resolution*.

Printer resolution   Measurement for the smallest point to be printed. Most laser printers have a resolution between 300 and 600 dpi (dots per inch). Phototypesetters have resolutions of 1200 to 2500 dpi.

Resolution   Measurement of the number of picture points per inch (dots per inch = dpi) or pixels per inch (pixels per inch = ppi) during printout or scanning. The maximum resolution depends on the resolution of the output device (printer) and the input device (scanner). The resolution gives the number of points per length unit. The higher the resolution the clearer and cleaner the image.

RIP (Raster Image Processor.) Specialized hardware or software (or a combination of the two) used to change text, format commands, image data and other page elements into lines of raster data. Part of typesetters and imagesetters.

Scaling Reducing or enlarging a scanned image.

Scan A scanned document in the memory of the computer or on disk.

Scan copy An original document (line art or halftone) which is to be scanned.

Scan file File in which a scan is saved in different file formats (for example, PCX, TIFF, EPS).

Scanner A device to convert analog information from a picture into a numeric value. During the scanning process the copy is scanned point by point by optical sensors. The brightness values are transferred into electrical signals used in the computer as picture points.

SCSI Small Computer Systems Interface. High bandwidth interface enabling peripheral devices such as scanners and hard disk drives to be attached to a computer.

TIFF (Tagged Image File Format.) An image file format developed by Aldus and Microsoft. Most scanning, picture editing and desktop publishing programs use this format for grey scale and line art images.

Vectorizing A file can contain a picture in its vectorized form, that is, it contains all the information necessary for creating the picture by means of coordinate references, describing origin, direction and length.

VGA (Video Graphics Array.) High resolution graphics standard for IBM PC systems and compatibles. The minimum for PCs supplied these days.

Index